BRIDGE MAINTENANCE, SAFETY AND MANAGEMENT

BRIDGE MAINTENANCE, SAFETY AND MANAGEMENT

By
Sanjeev Mathur

2012

SBS Publishers & Distributors Pvt. Ltd.
New Delhi

ISBN 13 : 9789380090474

First Published in 2012

Published by:

SBS PUBLISHERS & DISTRIBUTORS PVT. LTD.
2/9, Ground Floor, Ansari Road, Darya Ganj,
New Delhi - 110002,
INDIA
Tel: 0091.11.23289119 / 41563911 / 32945311
Email: mail@sbspublishers.com
www.sbspublishers.com

Printed in India by Chaman Enterprises, New Delhi.

Content

Preface

Maintenance of today's bridge infrastructure presents many challenges. Transportation engineering and maintenance personnel must maintain around the clock service to millions of people each year while maintaining millions of cubic meters of concrete distributed throughout their facilities. This infrastructure includes bridges. Presently only a limited number of accurate and economical techniques exist to test these structures for integrity and safety as well as insure that they meet original design specifications.

The bridge superstructure consists of that portion of the bridge above the bearings. Maintenance consists of repairs to the deck, handrail, curbs, floor system and structural members. Maintenance of structural members is normally minor unless members are damaged by collision or fire. Major damage to structural members which may affect the load carrying capacity must be reported immediately to Central Office Bridge Maintenance. If any doubt exists as to safety, the bridge should be closed until an inspection can be made and the damages determined. Girder ends under open expansion joints of all steel bridges should be flushed regularly to remove snow and ice control chemicals, dirt and debris. Truss members in the splash zone and lower chords should also be regularly flushed. Bridges with ASTM A709 weathering steel do not retain the protective oxide layer in the presence of snow and ice control chemicals. Therefore, it is important to regularly clean and flush these bridges under open joints.

Author

1

Highways Bridges Maintenance Strategies

INTRODUCTION

As bridges become older and maintenance costs become higher, transportation agencies are facing challenges related to implementation of optimal bridge management programmes based on life-cycle cost considerations. A reliability-based approach is necessary to find optimal solutions based on minimum expected life-cycle costs or maximum life-cycle benefits.

This is because many maintenance activities can be associated with significant costs, but their effects on bridge safety can be minor. In this paper, the programme of an investigation on optimum maintenance strategies for different bridge types is described. The end result of this investigation will be a general reliability-based framework to be used by the U.K. Highways Agency in order to plan optimal strategies for the maintenance of its bridge network so as to optimize whole-life costs.

As the existing stock of bridges continue to deteriorate, many countries, including the U.K., have to deal with the ever increasing demands on the limited resources available for their maintenance. In recent years, a number of bridge management systems have been developed with the purpose of prioritizing the necessary work. The first very comprehensive reliability-based bridge management system supported by the European Union is described in Thoft-Christensen.

The basic principle on which some of these systems have been based is that an optimum network level maintenance strategy can be determined by recording the present condition states of the bridges and their elements and then using deterioration prediction models related to different maintenance regimes.

However, the extent of bridge maintenance largely depends on the load carrying capacity of the bridges rather than on their condition alone. The implication is that estimates of maintenance needs should be based on bridge reliability rather than on condition states as defined in the current bridge management systems. Obviously, estimates of defects and deterioration are essential for determining bridge reliability.

In recent years there has been a search for including bridge reliability in the process of optimizing investments based on life-cycle costing.

Along these lines, the prime objective of bridge management is to determine and implement the best possible strategy that insures an adequate level of reliability at the lowest possible life-cycle costs or maximum life-cycle benefits.

The Highways Agency has to secure sufficient funds to enable it to maintain its structures in a safe condition. In order to justify these funds, the Agency needs to have an optimum strategy for the management of the trunk road network in England which includes some 16,000 structures most of which are bridges.

Although the number of structures is modest compared to the national stock of some 150,000 bridges, the truck roads in England carry one third of all traffic and more than half of all lorry journeys; as such, the maintenance of the structures on the network is of considerable national importance.

BRIDGE MAINTENANCE

A strategic plan was proposed by the Highways Agency in 1997 to determine its bridge maintenance needs for the future.

For a particular year, the strategic plan is intended to provide estimated levels of expenditure on both essential and preventative maintenance work. The justification for carrying out essential work is that, without it the element would be unsafe, and hence if the work cannot be carried out for some reason, in the interim period safety measures such as width or weight restriction have to be employed.

Such measures will cause traffic disruptions which can be estimated in terms of user delay costs. The justification for preventative work is that if it is not done at the time it will cost more at a later stage to keep the element from becoming critical. Also required as part of the overall maintenance regime is routine maintenance, which covers items such as inspections, drain cleaning and routine minor works.

In an ideal situation, the expenditure would be as shown in Fig. 1.1. Wallbank *et al.* If however, insufficient funding were provided each year, the amount of essential work required for structures to remain in service would start to increase as shown in Fig. 1.1. It is the purpose of the strategic long-term plan to identify the optimum expenditure profile.

Development of the strategic plan required the estimation of probability distributions for the maintenance intervals, preparation of typical maintenance costs, and application of the results to the range of bridge types and ages which make up the Highways Agency's bridge stock. These stages are described in Wallbank *et al.*

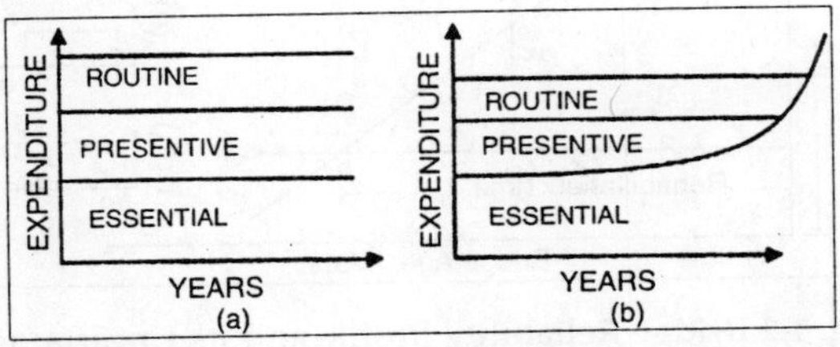

EXPENDITURE
ROUTINE
PRESENTIVE
ESSENTIAL
YEARS
(a)
EXPENDITURE
ROUTINE
PRESENTIVE
ESSENTIAL
YEARS
(b)

Fig. 1.1 Bridge Maintenance Programmes: (a) Ideal Bridge Maintenance Programme; and (b) Effect of Long Term Underfunding

One of the most important items required for the implementation of the strategic plan is the probabilistic distributions of the rates of rehabilitation or replacement of the various bridge types with and without preventative maintenance applied to them during their lifetime. Also required are the probabilistic rates of applying maintenance actions such as repainting of steelwork. The bridge rehabilitation rates can be determined using three methods.

The first, and the simplest, method is to base them on the expert opinions of experienced bridge engineers. This method was used by the Highways Agency for its first strategic plan in 1997.

The second possible method is to collect available data on rehabilitation or replacement work carried out by the maintaining authorities in the past. The third possible method for determining bridge rehabilitation rates is by using reliability-based studies of whole life performance under different maintenance regimes.

Bridge reliability analysis is essential for this purpose since there are many uncertainties in the lifetime process and these have to be dealt with in a rational manner. As shown in Figure 1.2 reliability level is affected by many uncertainties, including the 'as constructed' structural reliability, the damage initiation time, and the rate of reliability deterioration.

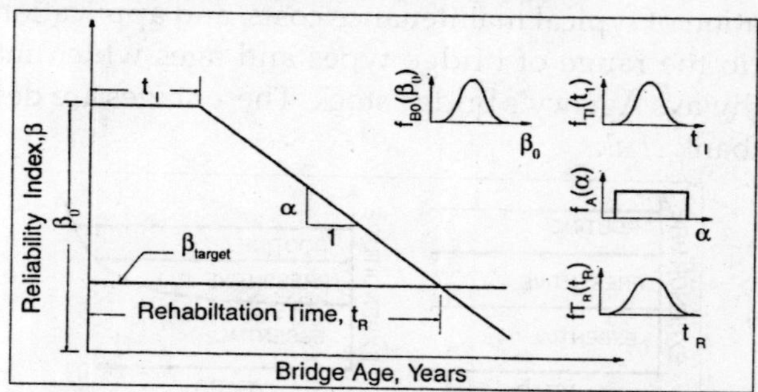

Fig. 1.2 Bridge Reliability Profile and its Uncertainties

The predicted performance curve for any group of bridges will, with time related deterioration, reach the assessment (minimum acceptable) level of performance at some point in the future and when that happens the bridges will have to be replaced or rehabilitated. The length of time from construction to the time of rehabilitation will obviously depend upon the reliability profile, which itself will depend upon the assumed maintenance regime. The probability distribution of that occurrence for the group is the required replacement/ rehabilitation rate for that bridge type.

Such probability distributions were recently obtained by Frangopol *et al.* Thoft-Christensen for steel/concrete composite and reinforced concrete bridges, respectively. A further type of uncertainty involves the average costs of rehabilitation and preventative actions which will be required to cost the different strategy options, and it will affect the final expenditure profiles.

The next step in developing the strategic plan is to, for each preventative maintenance scenario, multiply the numbers of bridges of each type constructed in a particular year, with the predicted rates of rehabilitation with and without (separately)preventative maintenance.

This will provide the numbers of bridges to be rehabilitated in any particular year in the future. Similarly, the numbers of bridges are also to be multiplied to the rates of preventative maintenance to obtain the numbers of bridges in any future year which will have the corresponding maintenance action carried out on them. Finally, the last stage is to choose the best maintenance strategy for each bridge type.

OPTIMUM MAINTENANCE STRATEGIES

As previously mentioned, a project was commissioned in 1998 by the Highways Agency to determine optimum maintenance strategies for different bridge types. This section briefly describes the tasks of this project and presents some results. The four major tasks of this project are data collection, development of strategies, probabilistic modelling, and reliability-based optimization.

DATA COLLECTION

It was originally intended to obtain data from 24 typical

bridges to reflect the main types 'reinforced concrete, pretensioned concrete, post-tensioned concrete and steel/ concrete composites, and different age groups. These bridges were identified by WS Atkins However; it was found that the Highways Agency's database did not contain appropriate data on maintenance history or present condition. Instead, unit costs were estimated for a series of maintenance options, based on data from current experience, as indicated in Table.

The triangular distributions for these costs have three parameters which are represented by an ordered triplet (a, b, c,) where a, b and c represent the minimum, mode and maximum values of cost respectively. In addition, the numbers of bridges of each type built in each year were identified, as shown in Table.

MAINTENANCE STRATEGIES

The maintenance strategies for each bridge type and age group have to be based on results from earlier research projects and operational experience.

The strategies include a "do nothing" strategy, involving no maintenance at all until repairs become essential. Also a "maximum maintenance" strategy has to be considered, whereby the bridge would receive frequent attention with the object of maintaining it in a pristine condition. The effect of essential maintenance is defined as the amount by which this activity improves the bridge reliability.

The effect of preventive maintenance is defined by the reduction in the rate of deterioration and, in some cases, by improvement in the bridge reliability.

A significant difference between essential and preventive, maintenaxnce is that essential maintenance is normally undertaken when the bridge reliability has fallen to, or below, the target value, whereas preventive maintenance is undertaken when the bridge reliability is still above the target value.

Figure 1.2 Frangopol *et al.* shows a comparison of present values of expected cumulative-time cost associated with three bridge maintenance strategies.

Table. Bridges Selected by WS Atkins

ID. Key	Description	Years of Const.	Capacity
	Pre 1955		
7914	1 span, Simply Supported beam and slab (cast iron /masonry) U/B	1929	3T ALL FEGp 2
7007	1 span Arched Masonry U/B	1840	40T ALL 33.8 HB
5549	3 span simply supported steel gdr &RC slab U/B	1926	40T ALL 45 HB
9455	2 span simply supported RC beam and slab U/B	1937	45HB
	1955-1964		
1594	3 span continuous comp-osite steel/RC U/B	1962	45 HB
966	5 span simply supported insitu PSC beams & RC slab U/B	1963	45 HB
5289	3 span simply supported Precast PSC beams and mass conc. slab U/B	1958	45 HB
12075	1 span simply supported composite steel RC deck U/B	1962	40T ALL 30 HB
	1965-1974		
14451	Instiu RC Solid Deck Slab and Arch	1966	40T ALL & 37.5 HB
1150	Continuous Precast post tens-ioned and Instiu RC box beam with cantilever wings	1970	HA + 37.5HB
1741	Insitu concrete slab on rolled steel beams continuous accross 6 spans	1966	HA + 45 HB

1174	Continuous voided Insitu RC solid slab	1966	40T ALL +45HB
	1975 - 1984		
15785	Continuous voided Institu PSC slab	1979	40T ALL or 45HB + AIL
15769	Continuous PSC voided slab	1975	HA
15767	Simply supported voided Insitu PSC slab with RC canatilever	1975	40T + 25HB
9507	Simply supported composite steel beam and RC slab	1979	HA + 45 HB
	1985 - 1994		
15974	3 span continuous composite steel/RC slab U/B	1990	HA + 45 HB
17996	1 span simply supported Insitu voided PSC deck U/B	1988	HA+ 45 HB
12335	2 span arched brick/masonry U/B	1985	HA
18900	1 span simply supported Pre-cast PSC and Insitu RC	1992	HA + 45 HB
	Post 1994		
N/A	7 span continuous steel U.B. composite with RC slab	1994	HA + 30 HB
N/A	Solid RC slab deck	1996	HA + 45 HB
N/A	Single span solid RC deck with integral abutments	1996	HA + 45 HB
N/A	Continuous 3 span solid RC deck with unbonded post tensioning	1996	HA + 45 HB

Strategy A consists of essential maintenance only (two essential maintenances) A-E1 and A-E2, are used during the life of the bridge, Strategy B consists of preventive maintenance only (five preventive maintenances, B-P1 to B-P5, are used during the life of the bridge,) and, finally Strategy C uses both essential maintenance, C-EI, and preventive maintenance, C-P1 and C-P2, during the bridge life-cycle.

Table. Estimated Maintenance Unit Costs

Maintenance Activity	Unit	Rate (pounds)			Frequency (years)		
Preventative Maintenance							
Painting	sqm	5	25	45	10	15	20
Deck Expansion Joints							
- maintenance	lin m	39	49	62	5	7.5	10
- Replacement	lin m	170	200	230	25	30	35
Waterproofing	sqm	16	23	34	25	30	35
Surfacing	sqm	15	22	32	25	30	35
Silane	sqm	15	20	25	10	15	20
Desalination							
Cathodic Protection							
- Temporary Support	lin m	6000	7000	8000			
- Concrete Repair	sqm	1800	2200	2600			
- Install CP System	sqm	150	200	250	20	30	40
- Maint. Of Anodes	sqm	70	100	130	7.5	10	12.5
- Inspection	sqm	20	30	40	7.5	10	12.5
Minor repairs	sqm	1300	1800	2300	10	15	20
Bearing Replacement	no.	1000	1500	2000	25	30	35
Essential Maintenance							
Concrete Repairs							
- Bridge Deck	sqm	500	900	1300			
- Crossbeam	sqm	500	650	800			
- Column	sqm	200	300	400			
- Abutments	sqm	600	1000	1400			
- Piers	sqm	300	700	1200			
Parapets							
- Upgrading	lin m	18	24	28			
- Replacement	lin m	68	90	119			
Replacement							
- Deck	sqm	100	200	300			
- Crossbeam	lin m		22000				
- Column	no		450000				
- Pier	sqm	1000	1300	1600			
- Tendon							

Note: Costs exclude VAT

Table. Bridge Stock by Year of Opening and Type

Year Opened	Reinforced Concrete		Post Tensioned		Pre Tensioned		Steel/Conc. Composite		Other		Total	
	UB	OB	UB	OB	UB	OB	UB	OB	UB	OB	UB	OB
<1955	262	2	1	1	6	0	107	3	345	2	721	8
1955	6	0	0	0	0	0	2	0	2	0	10	0
1956	15	1	0	2	0	1	5	0	1	0	21	4
1957	13	1	0	2	2	1	6	0	0	0	21	4
1958	21	0	0	3	1	1	6	2	0	0	28	6
1959	82	93	3	2	2	2	12	4	0	0	99	101
1960	62	18	9	10	8	7	16	7	0	0	95	42
1961	43	11	11	11	11	6	18	5	1	0	84	33
1962	96	47	21	12	11	11	29	29	0	0	157	99
1963	48	25	12	8	8	9	9	10	0	0	77	52
1964	37	9	8	14	12	20	18	8	0	0	75	51
1965	82	35	26	18	23	23	13	37	1	0	145	113
1966	102	25	27	5	12	12	18	8	0	0	159	50
1967	98	59	32	22	30	10	12	32	0	0	172	123
1968	60	12	18	13	25	19	29	25	1	0	133	69
1969	84	33	14	6	40	9	30	74	0	0	168	122
1970	122	78	99	20	53	54	19	53	0	0	293	205
1971	105	68	78	10	74	30	29	88	0	1	286	197
1972	56	32	27	4	41	25	10	29	0	0	134	90
1973	63	40	17	9	44	33	19	32	0	0	143	114
1974	103	81	30	19	48	49	20	11	0	0	201	160
1975	131	110	61	25	58	53	18	18	0	0	268	206
1976	85	61	13	10	38	33	16	10	0	1	152	115
1977	72	45	13	5	19	21	17	4	0	0	121	75
1978	57	40	31	1	24	48	17	11	1	0	130	100
1979	39	51	20	3	17	11	10	2	0	0	86	67
1980	38	44	6	4	28	37	9	9	0	0	81	94
1981	51	27	19	6	18	12	13	3	0	0	101	48
1982	53	41	10	5	13	13	14	8	0	0	90	67
1983	48	22	12	2	23	26	8	2	0	0	91	52
1984	43	56	10	1	16	18	3	5	0	0	72	80
1985	55	29	6	5	51	27	14	15	2	0	128	76
1986	42	53	8	1	26	13	12	10	0	0	88	77
1987	26	24	8	2	22	8	10	14	0	0	66	48
1988	52	38	10	5	19	7	10	6	0	0	91	56
1989	35	26	11	1	23	13	8	6	0	0	77	46
1990	55	60	35	4	25	3	27	23	1	0	143	90
1991	51	56	9	2	24	10	15	25	2	0	101	93
1992	48	58	8	1	21	3	18	13	0	1	95	76
1993	31	8	9	1	15	5	20	14	0	0	75	28
1994	22	13	9	0	2	6	5	18	2	0	40	37
1995	45	22	3	0	7	0	23	24	1	0	79	46
1996	10	3	0	1	4	4	11	13	0	0	25	21
1997	23	13	1	0	1	2	19	3	0	0	44	18
1998	0	0	0	0	1	0	0	0	0	0	1	0
Total	2672	1570	745	276	946	695	744	713	360	5	5467	3259

Note: UB = Underbridge, OB = Overbridge

The optimum maintenance strategy is obtained by choosing the least expensive present value of expected cumulative cost.

As shown, the optimum maintenance strategy is time dependent.

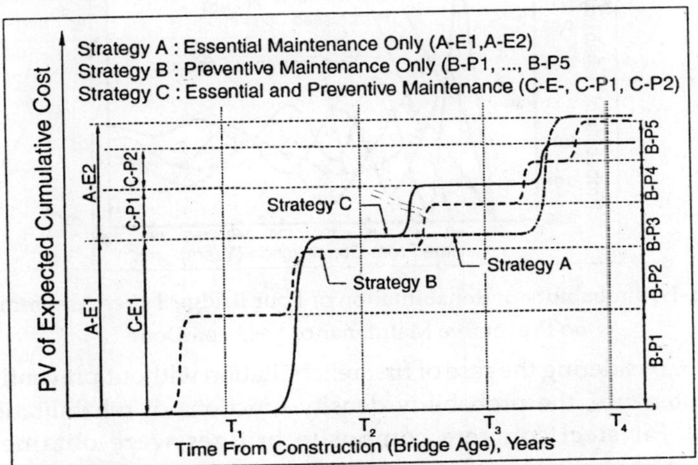

Fig. 1.3 Present Value of Expected Cumulative Costs for

PROBABILISTIC MODELLING

In order to find the optimal maintenance strategy for each bridge type, the present value of the expected cumulative cost of maintenance with and without preventive maintenance has to be obtained.

As a first step in this computation, the probability of rehabilitation has to be obtained. Figure 1.4 shows the probability of rehabilitation for four bridge types assuming no preventive maintenance has been done.

The computation of these probabilities is based on triangular distributions of rehabilitation rates predicted by experts for two different situations: (a) first rehabilitation assuming no preventive maintenance has been done; and (b) second rehabilitation assuming no preventive maintenance has been done.

As previously mentioned, a rational method for determining bridge rehabilitation rates is by using reliability-based studies of whole-life performance.

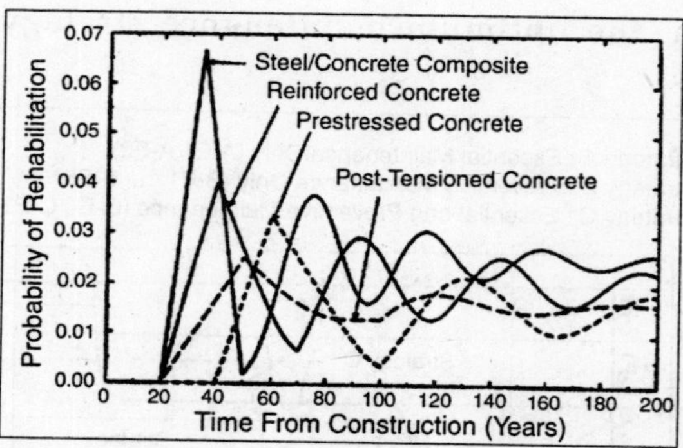

Fig. 1.4 Probability of Rehabilitation of Four Bridge Types Assuming no Preventive Maintenance has been Done

Considering the case of first rehabilitation without preventive maintenance, the probability density functions of rehabilitation rates for steel/concrete composite bridges were obtained. Figure 1.5 shows these functions assuming a target reliability level of 4.6. Research efforts are now in progress in Boulder and Aalborg to obtain the probability density functions of rehabilitation rates assuming preventive maintenance has been done.

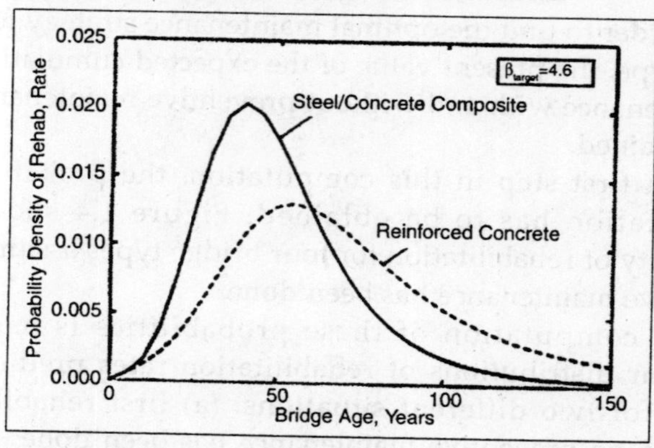

Fig. 1.5 Probability Densities of Rehabilitation Rates for Steel/Concrete Composite Bridges

RELIABILITY -BASED OPTIMIZATION

To implement the best reliability-based maintenance

strategy for each bridge type, the minimum expected life cycle cost solution has to be found. Whole-life costs have to be discounted using accepted rates. A considerable amount of sensitivity testing has to be undertaken, so that the effects of changed parameters on the optimum solution can be examined.

2

Bridge Maintenance

DESCRIPTION

Bridge construction and rehabilitation projects are complicated by the environmental sensitivities of working in riparian areas, including restricted work times to accommodate spawning periods of various aquatic species.

The majority of the steel bridges in the interstate highway system were constructed between 1950 and 1980; up until the mid 1970s, virtually all steel bridges were protected from corrosion by three to five thin coats of lead and chromate containing alkyd paints, creating dramatic complexity and cost increases for major and routine level bridge paint maintenance.

Lead-based paint abatement and the related issues of environmental, worker, and public protection came to the forefront of the maintenance painting industry in the mid-1980s to early 1990s when regulations on hazardous waste disposal and lead exposure to workers were promulgated by EPA and the Occupational Safety and Health Administration. The changes wrought by these rules still shape the direction of maintenance painting in the bridge industry.

NHI has developed a training course, "Hazardous Bridge Coatings: Design and Management of Maintenance and Removal Operations - NHI Course # 13069" for FHWA and State bridge engineers in the area of bridge coatings maintenance and specification. This course includes guidance on coatings selection, surface preparation specification, and environmental and worker safety issues and covers some of the information in a number of the following sections.

In addition to coatings and coating removals, bridge maintenance activities include repairing bent or damaged steel beams, cracked or spalled concrete, damaged expansion joints, and bent or damaged railings. These activities can entail operation of support vehicles and equipment, pavement repair, welding and grinding operations, and associated pollutants. Environmental stewardship practices under paving, structural pavement failure 'digouts,, pavement grinding, and concrete slab and spall repairs may be pertinent for bridge repairs.

PREVENTATIVE BRIDGE MAINTENANCE PRACTICES

Preventative bridge maintenance avoids larger scale work in stream environments, and thus makes sense from the standpoint of stewardship of both natural and financial resources. Preventive maintenance is defined as a planned strategy of cost-effective treatments applied at the proper time to preserve and extend the useful life of a bridge.

Bridge maintenance encompasses:

- Cleaning activities, including annual water flush of all decks, drains, bearings, joints, pier caps, abutment seats, concrete rails, and parapets each spring.
- Preventive maintenance activities such as painting, coating and sealant applications and for routine, minor deck patching and railing repairs.
- Technical and specialized repairs, including jacking up the structures, crack repairs, epoxy injection, repairing or adjusting bearing systems, repair and sealing of expansion joints, repair or reinforcement of main structural members to include stringers, beams, piers, pier and pile cap, abutments and footings, underwater repairs, major deck repairs, and major applications of coatings and sealants.
- Stream channel maintenance including debris removal, stabilizing banks and correcting erosion problems.

LIFE CYCLE DECISIONMAKING AND ACCOUNTING FOR ECOLOGICAL RISKS

Transportation Asset Management is driven by policy and performance and considers alternatives and trade-offs, evaluating competing projects and services based on cost-effectiveness and anticipated impact on system performance. As such it employs systematic, consistent business processes and decision criteria and makes good use of information and analytic procedures. In order to maintain and repair bridges within limited budgets, DOTs are establishing procedures for early detection of problems, timely repair, good preventative maintenance routines, and consideration of long term effectiveness of dollars spent.

NCHRP Report 483-Bridge Life Cycle Cost Analysis. Part one establishes guidelines and standardizes procedures for conducting life-cycle costing. Part two is the guidance manual for using the software to evaluate maintenance, repair, and new bridge alternatives. The AASHTO bridge management software, Pontis, can be used to: inventory elements of a bridge (such as coated steel girders); run deterioration and cost models to determine long term preservation policies; determine preservation needs and schedules; and provide network-level performance measures. European models have gone further in incorporating environmental aspects.

The LIFECON Life-Cycle Management System 'LMS, is a European model of a predictive and integrated LMS for concrete infrastructures, on both a short term and a long term basis, developed to facilitate decisionmaking. The system is divided into three levels of structural hierarchy: component and module, object, and network, with component- and module-level systems that address structural components such as beams and columns and their combinations in modules.

The object-level system deals with complete structures or buildings. The network-level system treats networks of objects such as stocks of bridges or buildings. Besides a structure's observed condition and evaluated urgency of repair, the life-cycle costs, user costs, minimum requirements of structural performance, structural risks, traffic and other operational requirements, aesthetics, environmental risks, and ecological pressures can be taken into account for multiple-attribute planning on all hierarchical levels of the system.

The life-cycle analysis and optimization module involves the data applications for studying the economy of the life cycle and cost-effectiveness of optional maintenance, repair, and rehabilitation strategies. Alternative strategies are compared as life-cycle activity profiles over a defined time frame. The purpose of life-cycle analyses is to find the optimal activity profiles to reach the targets. In January FHWA announced that Highway Bridge Replacement and Rehabilitation Programme 'HBRRP, funds can be used to perform preventive maintenance on highway bridges. Preventive maintenance activities eligible for funding include sealing or replacing leaking joints; applying deck overlays that will significantly increase the service life of the deck; painting the structural steel; and applying electrochemical chloride extraction treatments to decks and substructure elements.

FHWA is currently in the process of clarifying language and restructuring the National Bridge Inspection Standards. Proposed changes will reorganize the standards into a more logical sequence and make them easier to understand for inspectors, and state and federal highway administrators.

Bridge Inspection and " Smart Bridges" for Preventative Maintenance

In order to conserve fiscal and natural resources and ensure safety, DOTs are investing in bridge inspection for preventative maintenance and "smart bridges" that may forestall larger construction projects in and adjacent to streams. An increased emphasis on bridge safety and more rigorous inspection protocols followed the December 1967 collapse of a bridge over the Ohio River near Point Pleasant, WV, which claimed more than 50 motorists' lives.

As a result of the incident, Congress began to require the inspection and inventory of all bridges on the National Highway

System, with a specific provision that each bridge's load-carrying capacity be determined. More than 40 state DOTs and 100 engineering consulting firms now use the Bridge Rating and Analysis of Structural Systems (BRASS) software suite developed by the Wyoming DOT. Programmes within the BRASS suite now include applications specific to steel, timber and concrete girders, steel girder splices, piers, trusses, culverts, and poles, as well as for illuminated signs and signals.

The culvert design portion of the package comes from the North Carolina DOT; pre-stress girder design from Kansas; steel-field splice design from Nebraska; pier analysis and design programme work by the Portland Cement Association, Georgia DOT and Montana DOT; truss rating from New York; and, cantilever pole analysis and design from Louisiana. FHWA's Construction and Maintenance Fact Sheet on Bridge Preservation identifies best practices in bridge preventative maintenance, highlighting PennDOT's programme. PennDOT maintains the third largest number of State bridges in the Nation, spending $300 million on 250 bridge projects each year.

To keep costs down and ensure safety, PennDOT has found that it is vital to have both proper and frequent inspections and a good preventive maintenance programme. PennDOT's team of 50 bridge inspectors and numerous other consultant inspectors inspect all of the agency's bridges at least once every two years.

The bridge data is then stored in a management system, allowing engineers to prioritize the maintenance and rehabilitation needs and make sound decisions as to how to best take care of the bridge infrastructure. Handheld electronic data collection tools are augmenting such management systems and the efficiency of bridge inspection.

The Pennsylvania Turnpike Commission (PTC) had a consultant develop and use a handheld data collection system for the National Bridge Inspection Standards inspections for more than 800 bridges on the Pennsylvania Turnpike. Over a four year period of routine inspections, contrasting handwritten reports and electronic data collection, inspectors and the PTC found the electronic reporting system reduced costs, improved quality assurance and quality control, and provided easier access to inspection information.

Connecticut DOT has been using electronic monitoring systems to keep tabs on the condition of some of its bridges. Systems of linked sensors provide data on structural integrity and wear, and contribute to bridge life and stress assessment data. Portable and continuous systems have been installed on 11 bridges since 2002, allowing for early repair in sites that need it, and saving an estimated $2.7 million. Likewise, high-tech optical sensors embedded in concrete beams in a bridge in Las Cruces, N.M. relay information to New Mexico State University researchers about the performance of the bridge's design and materials, letting them track structural soundness as the bridge ages.

Recent research has concluded that truck weight is one of the most significant factors in the repair and replacement life of bridges. TRB Special Reports 225 and 227, Truck Weight Limits: Issues and Options and New Trucks for Greater Productivity and Less Road Wear: An Evaluation of the Turner Proposal, respectively, noted that trucks produce significant damage to highway bridges. A truck's gross weight, axle weights, and axle configuration directly affect the useful life of highway bridge superstructures. Damage typically occurs in the bridge deck and in the superstructure elements including floor beams and girders, diaphragms, joints, and bearings.

Bridge costs associated with increased truck weights are the result of the accelerated maintenance, rehabilitation, or replacement work that is required to keep structures at an acceptable level of service. While prestressed concrete I-girder bridges and modern steel-girder bridges could withstand a 20 per cent increase in truck weight, such an increase would reduce the remaining life in older steel-girder bridges by up to 42 per cent.

The "smartest" bridge to date, in terms of density of sensors, is under construction in Star City, West Virginia; it contains 770 sensors, 28 data-collection boxes, and a central data processing unit. WVDOT is counting on the investment to "help the state make smaller, less costly repairs while problems are still manageable," said Deputy Commissioner Norman Roush, as well as conserve resources that may be spent through overdesign.

Engineering data collection will be able to be correlated with continuous environmental data collection on-site. West Virginia existing "smart" structures have already yielded valuable information, such as that concrete slabs 20 feet long are prone to cracking, while those 15 feet long are not.

Small Bridge Maintenance Activities

Some of the bridge maintenance activities that provide the biggest benefit for the smallest level of investment generally include:

- Eliminating deck joints in old bridges
- Repairing or installing new expansion dams on bridge decks
- Repairing bridge decks
- Maintaining proper deck drainage
- Restoring or replacing bridge bearings
- Repairing or replacing bridge approach slabs
- Repairing bridge beam ends and beam bearing areas
- Bridge painting

Successful control of pollution from bridge maintenance and repair involves minimizing the potential sources of pollutants from the outset.

MAINTAINING DRAINAGE FROM BRIDGE DECKS

Effective bridge deck drainage is important because deck structure and reinforcing steel is susceptible to corrosion from deicing salts; moisture on bridge decks freezes before surface roadways, hydroplaning can occur more easily; and drainage occurs over environmentally sensitive areas. Bridge deck drainage is often less efficient than roadway sections because cross slopes are flatter, parapets collect large amounts of debris, and drainage inlets or typical bridge scuppers are less hydraulically efficient and more easily clogged by debris

Because of the difficulties in providing for and maintaining adequate deck drainage, the following practices should be used:

- Gutter flow from roadways should be intercepted before it reaches a bridge.
- Zero gradients and sag vertical curves should be avoided on bridges.
- Run-off from bridge decks should be collected immediately after it flows onto the subsequent roadway sections where larger grates and inlet structures can be used.

BRIDGE CLEANING

Bridge cleaning consists of cleaning all bridge components that are susceptible to dirt, debris, bird dropping and deicing salts. Drainage systems and components subject to dirt or bird droppings accumulation need to be cleaned regularly by hand tools, air blasting or preferably water flushing.

- Dust or any material that could be inhaled should be avoided by the use of a proper respirator.
- Other components such as bare concrete decks, pier caps, abutment seats, bearing systems, non-sealed or open expansion joints, joint drainage troughs, head walls, wing walls, select beam flanges, truss joints etc. should receive a thorough water flush every spring as a bare minimum.
- Personnel should become familiar with various types of bearing devices. Mechanical bearing devices should be lubricated after cleaning to prevent rusting and assist in their movement.
- Clearing of weeds, float debris, brush and overhanging limbs from the vicinity of the bridge should be performed according to best practices in channel maintenance.

AVOIDING AND MINIMIZING IMPACTS TO FISH AND WILDLIFE

The federal Endangered Species Act drives much planning for how to avoid impacts to species listed as threatened or endangered. Floodplains are protected by federal Executive Order 11988, the Rivers and Harbors Act, and the Clean Water Act. State laws and community expectations impose additional requirements. State DOTs are developing many creative approaches to meet these demands and their own environmental stewardship commitments.

DOT and state DOT-DNR cooperative initiatives to identify and develop standards and methods for improving fish passage are discussed on Culvert and Fish Passage Design Practices. Other enhancement practices for bridges are discussed in the following section,

IDENTIFYING OPPORTUNITIES TO AVOID AND MINIMIZE IMPACTS

Impact assessment is a standard part of the NEPA process, as well as compliance with other environmental laws. A few DOTs have taken this a step further by systematically identifying opportunities for avoidance, minimization, and/or environmental enhancement on projects or across classes of projects.

Oregon DOT's Bridge System Identification of Avoidance and Minimization Opportunities

Oregon DOT is preparing an Environmental Baseline Report for every bridge requiring improvement over the next 20 years, to inform design teams of all opportunities to avoid or minimize impacts. Environmental Performance Standards serve as a single, common set of terms, conditions and design targets that apply to all bridge projects and form the basis of programmatic or batched permits from multiple agencies. A Comprehensive Mitigation and Conservation Strategy (CMCS) that integrates wetlands mitigation with habitat conservation, through a series of regional banks.

SCHEDULING MAINTENANCE AND IMPROVEMENTS TO SPEND MINIMAL TIME IN SENSITIVE ENVIRONMENTS

As bridge owners, DOTs continually seek ways to build

more durable structures, safely, and in an environmentally sound fashion. In sensitive environments such as stream channels and floodplains, resource agencies often have an interest in confining construction to certain windows of time, to minimize impacts. There are times of the year when the effects of pollution from bridge maintenance and repair would cause the most damage and times when the damage would be minimal. The exact timing depends upon the site and the species involved.

Practices may include:

- Scheduling bridge maintenance to avoid egg incubation, juvenile rearing and downstream migration periods of fish.
- Calling upon state DOT fish and wildlife specialists or local fish and wildlife agencies for assistance in scheduling to avoid aquatic impacts.

For example, TxDOT has modified timing of maintenance modifications to protect bats in bridges, including postponing tree trimming and/or bridge maintenance work until outside of bat season. DOTs and resource agencies may find common ground in the interest in fast construction, if resources can be mobilized and the project completed in the time allowed.

To meet these needs and reduce the time to construct a bridge while maintaining or improving quality, prefabricated structures, more rapidly constructible details, and self-propelled modular bridge transporters are becoming more common.

USING PRE-FABRICATED BRIDGES TO HELP ACCOMMODATE STREAM/FISH TIMING RESTRICTIONS

Using prefabricated bridge elements and systems makes construction less disruptive for the environment. As traffic and environmental impacts are reduced, constructability is increased, and safety is improved because work is moved out of the right-of-way to a remote site, minimizing the need for lane closures, detours, and use of narrow lanes. Prefabrication of bridge elements and systems in a controlled environment without concern for job-site limitations can increase quality and lower costs, especially where use of sophisticated techniques would be needed for cast-in-place, such as in long water crossings or higher structures, like multi-level interchanges. NCHRP Synthesis 324 recently concluded that while prefabricated bridge components are more expensive in some cases, environmental impacts are reduced, quality is generally higher, and costs may fall as standardization increases in the industry.

Precast bridges consisting of pretensioned girders, posttensioned spliced girders, trapezoidal open-box girders, and other types of superstructure members are becoming more common due to their potential for accelerating construction and solving constructability issues in certain cases. Handling and shipping limitations often control the span capability of pretensioned girders. Spliced girder design has provided a solution where a one-piece pretensioned girder could not otherwise have been used.

Tennessee is developing bridge systems for steel bridges that can be erected similarly to precast, prestressed beams made continuous, with cranes of the same or lower lifting capacity, and can be fabricated at a reduced cost to be competitively priced.

In February 2003, FHWA and AASHTO sponsored a conference showcasing the successful uses and benefits of prefabricated bridge elements and systems. Along I-40 in Oklahoma where the cost of labour and materials to replace a collapsed bridge across the Arkansas River was $11.8 million, while the cost to control traffic and detours around the construction zone was $12 million, the higher upfront costs of prefabricated materials were offset by the amount of money saved by reduced construction times.

In San Juan, Puerto Rico, construction crews were able to minimize traffic delays and reduce construction times by using prefabricated bridge modules, while building four bridges at a congested intersection of a four-lane arterial. Crews completed the first bridge within 36 hours, and construction of each of the other three bridges took only 26 hours, using prefabricated elements. The conference profiled several new technologies and construction techniques, including new specifications for high-performance concrete and steel and self-compacting concrete, and contracting procedures that ensure that bridges are built within one day and charge high penalties for any delays.

Using prefabricated substructure elements reduces the amount of heavy equipment required and the amount of time required on-site for heavy equipment, causing less disruption to sensitive environments.

For example, the Hawaii DOT's Keaiwa Stream Bridge minimized environmental disruption because deck topping did not require shoring or falsework in the streambed, and minimized traffic disruption because precast planks were fabricated during pier construction. In North Carolina, to avoid placement of heavy equipment in a sensitive environment on the Blue Ridge Parkway, the Linn Cove Viaduct on the Blue Ridge Parkway was built in one direction from the south abutment to the north almost entirely from the top down.

The only exceptions to the top down method were construction of the initial span on falsework and construction of a temporary timber bridge that enabled the micropile foundation drilling machine to prepare several of the foundation sites ahead of the superstructure erection. Precasting each segment of the bridge allowed construction workers to assemble the bridge with little impact to the most environmentally sensitive section of Grandfather Mountain. This bridge proved that a design could be environmentally sensitive in addition to being utilitarian and economical.

The Wolf River Bridge in Fayette County, Tennessee, crosses sensitive wetlands and carries the only east-west route through its geographic region. TDOT designers selected precast prestressed beams to facilitate speedy construction and allowed optional stay-in-place precast prestressed concrete deck forms. TDOT and the contractor developed details for precasting bent caps in two pieces to suit staged construction. Construction of the 1,408-foot long, 46-foot wide bridge was completed in eleven months without putting any equipment in the surrounding wetlands. Photo Credits: Tennessee Department of Transportation. TxDOT developed two new bridge superstructure systems that have maximum span lengths of 115 ft and a total superstructure depth of only 38 in. and are totally prefabricated: a steel tub girder and a prestressed concrete pretopped U-beam.

The steel tub-girder system uses a conventional prefabricated trapezoidal steel girder, which is topped by a concrete slab before transport to the bridge site. To achieve the shallow superstructure depth of 38 in., shoring the beams during slab placement makes them composite for all loads. After slab placement, the beam is hauled to the bridge site and erected on the bridge piers. A simple cast-in-place closure pour joins the deck girder sections after they are in place.

The prestressed concrete pretopped U-beams use a portion of the existing Texas U-beam form system. Each beam is fabricated as a closed U-beam and hauled to the contractor's yard, where a 4-in. topping is placed before beam erection. A cast-in-place closure pour joins the deck girder sections after erection. Texas DOT anticipates that these two systems will be used over the next 10 years for the rapid construction of nearly 150 bridges that cross I-35 in central Texas.

Construction of the first four such structures begins in spring 2005. It is expected that girder erection and closure-pour placement will take less than 24 hours and that bridges will open to traffic after as few as 4 days. In WSDOT is beginning research on Precast Systems for Rapid Construction of Bridgesthat will build on TxDOT's experience and experimentally verify that precast systems can be constructed to pass rigorous seismic standards.

FHWA's segmental concrete bridge technology offers resources and best practices in SCBT bridge use, a method of joining multiple cast-in-place or precast bridge elements to form a continuous span. Precast segmental concrete construction is especially useful and efficient in difficult construction sites, which have included both urban and natural areas. FHWA's addresses engineering issues and construction methods, and features a photo gallery and an archive of "Ask the Experts" questions that have been submitted by site users and answered by team members. Geosynthetic Reinforced Soil (GRS) abutments and walls can help reduce the access needed for large equipment.

Self-Propelled Modular Bridge Transporters

Prior to the use of self-propelled modular transporters (SPMTs), rapid bridge replacement techniques were often limited to prefabricated bridge components assembled in the field. SPMTs now allow erection of entire bridge superstructures, which can then be replaced quickly within timing restrictions and without cranes, temporary or extensive detours, or traffic delays. In addition, construction of bridges in a controlled environment can maximize the quality of the finished bridge.

Hydraulic, self-propelled platform trailers, supporting structures, jacking systems, barges, and project-built solutions, bridges with a weight of up to 6,000 metric tons (6,000 tonnes) have been moved transversely and longitudinally into precise position. The method is becoming increasingly common in Europe, for both highway and railway bridges.

An international scan was conducted in April 2004 to learn how other countries are using prefabricated bridge components to minimize traffic disruption, improve work-zone safety, minimize environmental impact, improve constructability, improve quality, and lower life-cycle costs. The top implementation recommendation from the scan team was the use of self-propelled modular transporters to move bridges into position in hours rather than the typical months required for conventional bridge construction.

REDUCING THE SPACE NEEDED FOR LARGE EQUIPMENT ACCESS

Geosynthetic Reinforced Soil (GRS) abutments and walls can reduce the space needed for access by large equipment in building bridge abutments. FHWA has completed a substantial amount of research on this technology and feels it has great potential for future application geotechnical programmes,

"Performance Test for Geosynthetic-Reinforced Soil Including Effects of Preloading, FHWA-RD-01-018" and "Effects of Geosynthetic Reinforcement Spacing on the Behaviour of Mechanically Stabilized Earth Walls, FHWA-RD-03-048.

ENHANCEMENTS TO BRIDGES AND STREAM ACCESS

In the course of bridge maintenance and improvements, a number of DOTs are exploring ways to improve the environment for people and animals. While the Design stage is key to building in improvements such as bridge extensions, for wider floodplain and stream channel protection, as Florida DOT considers as a matter of policy, much can be done to enhance the environment in the course of maintenance.

FISHING ACCESS

DOT maintenance staff usually know where the public is trying to access fishing sites and where such access is presenting a safety problem on the roadside and/or improvements could easily be made that would benefit the traveling public, with minimal investment. NYSDOT Niagara County maintenance staff took the lead in forming a partnership with local business, the NYS Office of Parks Recreation and Historic Preservation and the New York State Department of Environment Control, to provide a public fishing access site and picnic area at a popular salmon and trout stream—Keg Creek.

Prior to the improvement, anglers parked along the state highway and made their way down a very steep, slippery and dangerous ravine to fish for Lake Ontario's world famous migrating trout and salmon. NYSDOT maintenance crews designed and constructed a paved parking area, a series of wooden stairs and a picnic area with lumber donated by a local company and tables donated by the State Parks Department.

Bird and Bat Roosts in Bridges

Nests in and on bridges are often regulated under the Migratory Bird Treaty Act. Migratory bird concerns often require DOTs to time projects for "off- season" work so as not to disturb active nests. In some cases, large unoccupied nests may be relocated during the off-season. Maintenance forces have used deterrents for nesting swallows on bridges such as netting or discouraging coatings.

Virginia Department of Transportation Programme Key in Comeback of Peregrine Falcons

Peregrine falcons were classified nationwide as an endangered species from 1970-1999. In the 1980s, the falcons began showing a preference for VDOT's coastal bridges. At that time, VDOT began working with the Virginia Department of Game and Inland Fisheries and the College of William and Mary to aid bridge-nesting efforts. VDOT installed nesting boxes on ten bridges and video cable on one. From 1978 through 1985, the Department of Game and Inland Fisheries led a recovery effort that released 115 young birds on the Eastern Shore.

Rather than moving to western Virginia as anticipated, VDOT bridges became one of the most popular nesting sites. Native to Virginia's Allegheny and Blue Ridge mountains, peregrine falcons were nearly wiped out by unintended side effects of pesticide use by the early 1960s. All known breeding pairs east of the Mississippi had disappeared and their numbers were drastically reduced worldwide by the mid-90s. The Department reports falcon activity to wildlife experts, and limits maintenance work to avoid disturbing nesting pairs or their young.

In 1998 VDOT's Environmental Division earned a Federal Highway Administration Award for Excellence in Highway Design for Environmental Protection and Enhancement for the effort. VDOT employees have been recognized by the Board of Directors of Virginia Game and Inland Fisheries. VDOT has continued its habitat enhancement and sensitive maintenance practices even after the species was delisted. In the spring of 2001, VDOT and eight other public and private agencies began FalconTrak, a three-year programme to protect eggs and hatchlings, track young falcons via satellite, and monitor nests with video cameras for researchers and the public to view on VDOT's web site, Pairs of peregrine falcons are currently nesting on eight VDOT bridges and offspring are thriving.

Protecting and Increasing Bat Roost Habitat in Bridges

Bats are primary predators of vast numbers of insect pests that are extremely costly to farmers and foresters. One bat can easily eat 20 female corn earworm moths in a night, and each moth can lay as many as 500 eggs, potentially producing 10,000 crop-damaging caterpillars. Yet as few as eight caterpillars per 100 plants can force a farmer to apply pesticides.

A number of bat species nationwide are listed as threatened or endangered under the federal Endangered Species Act. Bats are especially susceptible to extinction because most species form large colonies in vulnerable locations, such as caves that are sometimes inadvertently sealed. In addition, bats usually produce only one pup per year. As a consequence of losing natural roosts in caves and old growth forest snags, bridges and culverts have become havens of last resort. Bridges from Canada to Florida are being used by at least 24 of the 46 North American bat species; it is estimated that within the southern United States alone, 3,600 highway structures are being used by approximately 33 million bats.

State DOTs can contribute to bat recovery at little or no cost, through proactive measures. Most bat species that will roost in bridges choose concrete crevices that are sealed at the top, at least six to 12 inches deep. 5 to 1.25 inches wide, and ten feet or more above the ground, typically not located over busy roadways. Day roosts are places that protect bats from predators and buffer weather conditions while resting or rearing their young. Such roosts are usually in expansion joints or other crevices. In contrast, night roosts, where bats gather to digest their food between nightly feeding bouts, are often found in open areas between bridge support beams that are protected from the wind. Retrofitting existing bridges and culverts proved highly successful in attracting bats, especially where bats were already using them at night.

Fig. 2.1 TxDOT Bats in Bridge Retrofit Partnership

Citizens have gotten involved as well. When 33,000 Mexican free-tailed bats became a nuisance in the school attic at Canadian Middle School in Canadian, Texas, teachers and students purchased materials, constructed, and installed alternate roosts for up to 50,000 bats in a nearby highway bridge by collaborating with Bat Conservation International (BCI) the TxDOT and local businesses.

Bats have the largest surface area to body mass of any mammal, and this requires greater energy to maintain body temperatures. Sun-warmed bridges help adult bats to conserve energy and foster development of their young. During the summer months, sun-exposed bridges act as thermal sinks, often achieving and holding temperatures above the ambient average for most of the 24-hour cycle.

Bat Conservation International cooperated with 20 state DOTs in a national study of bats in bridges and found 217 highway structures used as day roosts and 714 highway structures used as night roosts. Information from this study is summarized in this section and those following, pertaining to construction and retrofit recommendations. The study found that the higher, more consistent bridge temperatures are especially important in mountainous or desert regions where ambient temperatures fluctuate dramatically within a 24-hour cycle. An Oregon study found that bats prefer bridges with greatest sun exposures. Bridges receiving no sun had little or no bat use. This preference was especially obvious within partially shaded bridges, where roosting activities occurred only in the sun-exposed halves of bridges. The northernmost day roost discovered in this study was occupied by a maternity colony of roughly 300 little brown myotis in an Idaho bridge at 44 north latitude. However, the number of day roosts appears to drop rapidly above 42 north latitude.

Bats use parallel box beam bridges as day roosts more than any other kind. The next most preferred bridges are cast in place or made of prestressed concrete girder spans. These designs are the most likely to contain spaces suitable for bats. Although parallel box beam bridges were rarely encountered during the survey, they can provide numerous crevices of suitable width. Metal and small concrete culverts are the most frequently encountered highway structures and are the least preferred as roosts. Even ideal structures were rarely used by bats in areas dominated by open plains, perhaps due to a lack of appropriate habitat.

Creation of day-roost habitat for bats in new or existing highway structures is easy, often at little or no extra cost to the taxpayer. For new structures, the minimum needs for day-roosting bats can be met by specifying the proper dimensions for crevices such as expansion joints. Night use of highway structures is even more common; 29 per cent of all structures surveyed had signs of night-roost activity. In some regions of the southwest, all suitable structures were used by night-roosting bats. Night-roosting bats are believed to be attracted to bridges that provide protected roosts and have a large thermal mass that remains warm at night.

Bridges constructed of prestressed concrete girder spans, cast-in-place spans, or steel I-beams are preferred. Vertical concrete surfaces located between beams provide ideal protection from wind and are especially used when they are heated by full sun exposure.

Bats typically do not use bridges with flat bottomed surfaces that lack inter-beam spaces. They will avoid small culverts but will roost at night in the long concrete box culverts that often pass under divided highways, if the culverts are at least 5 feet (1.5m) tall. Night roosts appear to play important roles in body temperature regulation and social behaviour.

Fig. 2.2 Night Roosts Located in Open Spaces between Bridge Beams. Credit: Bat Conservation International

TxDOT Study and Bridge Modifications to Support Bat Usage

TxDOT and Bridge Engineer, Mark Bloschock, have received an award of excellence from Bats Conservation International, the first given to a person outside the field of wildlife conservation. Bloschock began a study with BCI in 1994 after a large colony of Mexican Free-Tailed bats settled under the Congress Avenue bridge in Austin and each wanted to determine why the bats settled there, whether bats might damage the bridge, and if there might be potential effects on human health.

The study determined that slot-shaped crevices under the bridge were similar in size to spaces found in bat caves and uncovered bat roosting preferences in both bridges and culverts throughout the state. The study indicates that minor modifications to highway structures can maximize or minimize the potential for use by bats, and that less than 0.01 per cent of Texas highway structures currently meet the day-roosting requirements for bats. In central, southern, and western Texas, there is a 62 per cent chance that structures with suitable characteristics will be used by bats.

The Mexican free-tailed bat was found to be the most frequent bat species day roosting in highway structures. Bridge characteristics preferred by day-roosting bats were defined by a paired comparison study where bat-occupied and unoccupied bridge characteristics were statistically compared. Today, 1.5 million Mexican Free-Tailed bats migrate from Mexico every year and stay from March to October under the popular Austin bridge. It's estimated the bats eat ten to 15 tons of insects on their nightly flights.

The sight of the bats taking flight at dusk from beneath the Congress Avenue bridge had another unexpected benefit as thousands of tourists visit the bridge to see the nightly flights. Each year, the bridge attracts tens of thousands of tourists from all over the world, and has been estimated to generate more than 8 million for the local economy. Based on the success of the bat habitat in Austin, Bloschock established the Bats and Bridges programme, which has spread to 24 states and 17 countries.

Overall, TxDOT has 218 structures currently used as roosts, almost three times as many habitats for bats as any other state taking part in the programme.

Practices to Incorporate in Design, Construction, and Retrofitting of Bridges for Bats

TxDOT performed a statewide evaluation used to identify the distribution of highway structures used by bats. Day-roosting bats prefer concrete bridges and culverts with secluded locations such as crevices that are 0.5 to 1.25 inches-wide (1.2 to 3.2 cm) especially those that are 12 inches deep (30 cm), have covered tops, and are located in central, southern, and western Texas.

Additional experiments further supported the results of the paired comparison and statewide evaluations. Bat colonies, even large ones, do not damage highway structures and water sources under roosts are not negatively impacted. Human health risks are minimized by educating people not to handle bats.

In sum, the study found that:

- Highway structures that incorporate crevices between 0.5 and 1.25 inches wide (1.2 and 3.2 cm) can provide ideal roosting habitat for several of the most rapidly declining and valuable bat species in Texas, especially if these crevices are 12 inches (30 cm) or more in depth and covered at the top.

- Bats typically use only concrete or wooden roosting surfaces, preferring the highest, darkest locations.
- Structures can be retrofitted with Bat-Abodes or concrete panels to create bat habitat.
- Large concrete culverts under divided highways can provide excellent roosts for threatened and endangered bats, though most would require provision of roughened ceiling cavities during construction.
- No structural damage, aquatic pollution, or disease transmission to humans has been associated with even the largest bat colonies living in Texas bridges and culverts, but warnings not to handle downed individuals or inhale dust associated with bird or bats droppings are recommended.
- Where bats are unwanted, simple elimination of preferred crevice widths can prevent potential nuisance problems.

Incorporating characteristics into new structures specifically for bats can be relatively inexpensive and easy to do. TxDOT has developed a bat-friendly domed culvert, for which customization costs are minimal; modifications can even be implemented during construction. As part of their national study, Bat Conservation International determined in consultation with state DOTs that bat-friendly habitat can be provided in either new or existing bridges or culverts, at little or no extra cost to taxpayers.

During construction planning, there are no costs for an engineer to specify the appropriate crevice widths of 3/4 to 1-inches (1.9 to 2.5 cm) for expansion joints or other crevices. Existing structures can be retrofitted with bat-friendly habitats using the designs described in the following sections. Signs of bat use in nearby bridges and culverts increase the chances of success for habitat enhancement projects.

Ideal day roost characteristics for crevice-dwelling bat species that use highway structures, include (in descending priority):

- Location in relatively warm areas, primarily in southern half of the country
- Construction material: concrete
- Vertical crevices: 0.5 to 1.25 inches (0.25 to 3 cm) wide
- Vertical crevices 12 inches (30 cm) or greater in depth
- Roost height: ten feet 'three meters, or more above the ground
- Rainwater-sealed at the top
- Full sun exposure of the structure
- Not situated over busy roadways

Culverts:

- Location in relatively warm areas
- Concrete box culverts
- Between five and ten feet (1.5 and 3 meters) tall and 300 feet (100 meters) or more long
- Openings protected from high winds
- Not susceptible to flooding
- Inner areas relatively dark with roughened walls or ceilings
- Crevices, imperfections, or swallow nests

The Texas Bat-Abode, Big-eared Bat-Abode, and the Oregon Bridge Wedge bat roosts are designed for day-roosting bats in bridges and culverts. In the protected environment of a bridge or culvert, a properly constructed and installed bat habitat made of quality materials should last as long as the highway structure. BCI would appreciate photographs of the installation and especially of bats using any bat-friendly modifications. The designs to specific bridges, or to report occupied units, contact Bat Conservation International (BCI), Inc., at 512/327-9721. BCI maintains a list of bats, documented bridge/culvert use, potential use, roost type (crevices or open beams), preference, nationwide distribution, and status.

Texas Bat-Abode

The Texas Bat-Abode is designed to retrofit bridges with bat habitat for crevice-dwelling species. It has an external panel on either side and 1x2-inch (2.5 to 5.1 cm) wooden spacers sandwiched between 0.5 to 0.75 inch '1.2 to 1.9 cm, plywood partitions. Recycled highway signs are ideal construction materials. Note that only the external panels need to be cut to fit the bridges' inter-beam spaces. The internal partitions should provide crevices 0.75 inch (1.9 cm) wide and at least 12 inches (31 cm) deep. Smooth roost surfaces need to be textured to provide footholds for bats on at least one side of each plywood partition (preferably both), creating irregularities at least every 1/8 inch (0.3 cm). Many methods have been tested to create footholds, such as,

- Using rough-sided paneling
- Coating the panel with a thick layer of exterior polyurethane or epoxy paint sprinkled with rough grit
- Attaching plastic mesh with silicone caulk or rust-resistant staples
- Mechanically scarifying the wood with a sharp object such as a utility knife
- Lightly grooving the wood with a saw 'do not penetrate to the first plywood glue layer,
- Lightly sandblasting the wood with rough-grit

The use of rough-sided paneling or polyurethane sprinkled with grit have provided the longest lasting results. Rust resistant wood screws should be used to assemble the spacers and partitions.

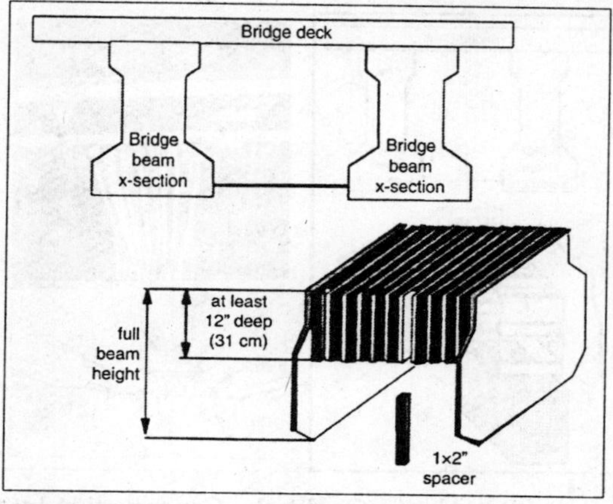

Fig. 2.3 Texas Bat-Abode for Crevice-dwelling Species.

Credit: Bat Conservation International

The Texas Bat-Abode should be installed in bridges that are at least ten feet above ground, free of vegetation, and not susceptible to flooding or easy vandalism. Measurements of the exact location where the Bat-Abode is to be placed will ensure a proper fit. The number of partitions is arbitrary and limited only by availability of materials and support for the weight of the Abodes. Because of the weight, it may be easiest to assemble the cut pieces in the bridge. In wooden bridges, the unit should be anchored to the structure with heavy-duty rust-resistant lag-bolts.

Big-eared Bat-Abode

Big-eared bats are frequent bridge users in both the eastern and western United States. They prefer open roost areas such as cave entry rooms, large hollow trees, darkened undisturbed rooms in abandoned houses, or between the darkened beams of quiet bridges over streams. The Big-eared Bat-Abode creates these conditions. The Big-eared Bat-Abode has two external panels with 1x2-inch spacers that are used as braces to hold the panels together with a plastic mesh lining to provide footholds for bats. The netting should be attached using rust-resistant staples. The other methods of creating footholds mentioned above would also be effective.

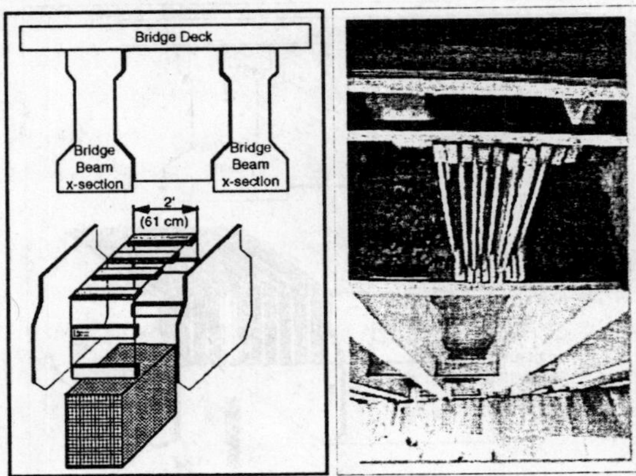

Fig. 2.4 Big-eared Bat-Abode. Credit: Bat Conservation International.

Several designs of the Texas Bat-Abode, such as this one modified for a steel I-beam bridge, have been used to attract thousands of bats. Complements of TxDOT and BCI. It may be easier to partially assemble the structure on the ground leaving one end panel off until it is placed in its chosen location. Units installed in wooden bridges can be anchored using heavy-duty rust-resistance lag bolts.

Because big-eared bats are very sensitive to disturbance, units should be placed in areas of low activity and painted a colour that does not attract attention. Big-eared bats are often found in low bridges darkened by thick vegetation growing along the sides. The Big-eared Bat-Abode should be placed at least six to ten feet 'two to three meters, above the ground in a secluded portion of the bridge. However, access to the fly-way entrance should not be blocked. Other bat species are likely to use this structure.

The Oregon Wedge

The Oregon Wedge is an inexpensive method of retrofitting bridges or culverts with day-roost habitat for bats. The Wedge is made from an 0.5 to 0.75 inch (1.2 to 2 cm) exterior grade plywood panel that is at least 18 inches high and 24 inches wide (46×61 cm) with three 1 × 2 inch (2.5×5 cm) wood strips attached along the top and sides, leaving an opening along the bottom. If larger panel sizes are used, vertical wooden pieces should be placed every 24 inches (61 cm) to support the plywood and prevent warping. The pieces should not run from the top to the bottom so that bats can move about within the panel.

The Wedge can be attached to a vertical4 concrete portion of a bridge or culvert using concrete anchor-bolts or a fast-drying environmentally safe epoxy cement (such as 3M Scotch coat 3-12). The transportation department should install the panels if anchor bolts are used.

If the panel is to be attached to wood, then use appropriate rust resistant wood screws. Before applying the epoxy, check the preferred installation site to make sure the support strips fit flat against the concrete surface.

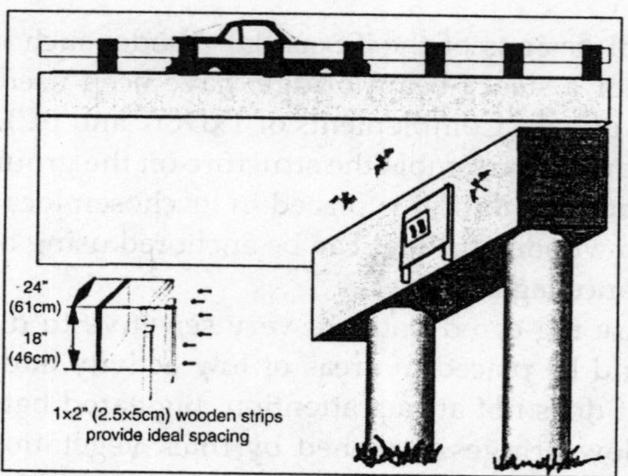

Fig 2.5: Designs Courtesy of David Clayton and Dr. Steve Cross and Bat Conservation Int'l

Wedge placement is possible on any adequately sized, flat concrete or wood surface. However, we recommend that the panels be placed near the sun-warmed road slab preferably as high as possible between heat-trapping bridge beams. They should be at least ten feet (three meters) above ground, with a clear flyway (at least ten feet), and be out of view or reach of vandals. The Wedge can be installed in the middle sections of culverts higher than five feet (1.5 cm). A Wedge should not be placed in structures that flood. As a precaution against flooding, a 1.5 inch (3.8 cm) gap can be left at each corner where the support strips join to act as an escape route in the event of fast-rising water.

Bat-domed Culverts

Fig. 2.6 Bat-domed Culvert. Graphics Courtesy of TxDOT and BCI.

The Bat-domed culvert is a modified concrete box culvert designed to accommodate large colonies of bats.

The dome has several bat-friendly characteristics:

- The height is increased
- Warm air is trapped
- Light intensity is reduced
- Air movement is reduced

Bat-domed culverts should be at least 5 feet (1.5 m) in height with an additional 1 to 2-foot (0.6 meter) raised portion centered in the culvert. The raised area can be any length from 2 to 50 feet, depending on the colony size preferred. The walls and ceilings of the raised area should be roughened to provide footholds for bats. The following method was used to produce suitable wall and ceiling textures. Using a crowbar, thin strips were removed from the surface of recycled plywood. The resulting roughened wood was then used as the form for pouring the concrete, which produced the desired textured surface within the domed area of the culvert. In addition, a method of attaching panels or partitions, such as female threaded inserts, can be incorporated into the raised walls and ceiling to create more surface area once the culvert is completed.

Bat-domed culverts should not be placed in areas susceptible to flooding. However, in the event of rising water, the dome may serve as a temporary air-trap. Almost any cave-dwelling species may use these, including several that are endangered.

Success in Retrofitting Bridges to Accommodate Bats

Retrofitting habitat into existing highway structures has become a popular and successful method of accommodating bats. Pre-surveys to look for bat signs in nearby bridges are useful to predict the success of proposed enhancement projects. Four bridges in Oregon and five bridges and two culverts in Texas with signs of night roosting were retrofitted with ideal crevices, and all were occupied by bats within the first year. All retrofit designs tested in bridges and culverts so far have successfully attracted bats, and at least six states are already using retrofitting projects to accommodate bats. Retrofits are adaptable to almost any structure and can be placed where they will have a high potential for success and that will minimize disturbance from maintenance or vandalism. Retrofits are generally inexpensive and can be sized to accommodate small or large colonies, with potential expansion by adding more units if initial efforts are successful. Retrofits are usually highly beneficial to local agricultural and can be easily moved if necessary.

Fig. 2.7 TxDOT's Concrete Version of the Oregon Wedge

Two basic designs can be used to retrofit almost any bridge or culvert. Texas Bat-Abodes can accommodate thousands of bats each, and have been modified to fit three different bridge designs. Four of the five tested were fully occupied, one within the first month.

The Oregon Wedge can house several hundred bats and has been accepted for day roosting by 12 species, including a maternity colony of Yuma Myotis. This design has been successful in both bridges and culverts in Oregon, Arizona, and Texas. The Texas Department of Transportation developed a concrete version that attracted bats within a year.

Preserving Portions of Old Bridges as Habitat

When old bridges must be replaced, some of those occupied by bats have been retained as wildlife sanctuaries. The Santa Barbara Public Works Department and Caltrans are collaborating to preserve a colony of 10,000 Mexican free-tailed bats and 200 pallid bats by retaining a portion of an old bridge that is surrounded by agricultural fields. It is calculated that these bats consume roughly 10,000 pounds (4,540 kg) of insects each summer, many of which are pests. Bridge habitat enhancement techniques are being developed in other countries. In Australia, the roost portion of an old wooden bridge was retained and incorporated into the underbelly of a new replacement bridge.

Further DOT Efforts to Identify Characteristics and Design Features of Roost Bridges and Conservation Efforts

FDOT is in the process of undertaking a survey of bridges with bats to help FDOT predict and control where bats will roost. At least five species of bats in Florida use concrete highway bridges as roosting sites. Because many natural roosts such as caves and large hollow trees are rare, bridges serve as the most common or primary roosting sites for bats in some areas.

The objectives of this project are to: (1) identify FDOT maintained bridges in Florida that are occupied by roosting bats; (2) summarize characteristics and design features of roost bridges, and correlate bridge features with presence, number, and species of roosting bats; (3) prepare guidelines for FDOT employees to record the presence of bats during routine activities; and (4) identify all bridges that support bat roosts and are planned for replacement by 2020, and to identify ways to conserve the roosts when these bridges are replaced.

Likewise Georgia DOT will (1) identify the highway bridges and select culverts in Georgia that are occupied by roosting bats; (2) evaluate the characteristics and features of bridges being used as roosts, including an assessment of surrounding habitat features; (3) recommend bridge design elements that provide the roost features preferred by bats; and (4) prepare standard procedures for assessing the presence of bats, minimizing disturbance to the bats, and preserving the existing or potential roosting opportunities during management, repair, and demolition of highway bridges by the Georgia Department of Transportation (GDOT). Those research results will be available in mid-2005.

BRIDGE PAINTING/COATING/SEALING AND CONTAINMENT STEWARDSHIP PRACTICES

Bridge painting/coating/sealing covers all protective and preventative maintenance activities designed to prevent deterioration of structure components. Components made of non-weathering steel are generally painted with a multicoat paint system to protect the steel from rust and corrosion. Bridges painted prior to 1975 typically used lead, chromium, or cadmium pigmented paints, which if removed must be removed according to strict EPA and OSHA guidelines and disposed of as a hazardous waste.

The 1990s saw great increases in the costs and complexity associated with steel bridge maintenance painting. While a low-tech, least cost approach to combating corrosion and deteriorating aesthetics prevailed as recently as 15 years ago, the increasing age of the infrastructure, increasing needs for immediate maintenance, over the past few years, environmental regulations have become the single most influential force affecting the bridge painting industry; specifically, the regulations regarding the VOC content of protective coatings and the environmental and worker health and safety regulations associated with the removal of lead-containing paint have had a significant impact on the bridge painting industry.

Table. Regulations Impacting the Bridge Painting Industry

Impacting Regulation	Effect on Coating Operations
OSHA; CFR 29 1926.62, Lead in Construction	Establishes guidelines for protection and monitoring of workers removing lead paint from bridges. Requires lead training and monitoring for workers.
EPA; Resource Conservation and Recovery Act (RCRA)	Regulates the handling storage, and disposal of lead and other heavy metals, cont -aining waste. Can increase the cost of disposal of waste

	from bridge paint removal by 10 times.
EPA; Comprehensive Environmental Response Compensation and Liability Act 'CERCLA or Superfund,	Assigns ownership of and responsibility for hazardous waste to the generator "into perpetuity."
EPA; Clean Water Act	Regulates discharge of materials into waterways.
EPA; Clean Air Act Amendments	Mandates restrictions on allowable volatile-organic-compound (VOC) content of paints and coatings. Regulates discharge of dust into air from bridge painting.

The impact of regulatory compliance has spurred a shift in focus to achieving long-term effectiveness for the dollars spent, both to maximize the return on DOT investment and minimize environmental impacts. A number of FHWA research projects with regard to bridge coatings have addressed the life-cycle cost issue in detail. Research on the removal of lead-containing paint considered the relative cost increases associated with changing regulations that deal with removal and handling of hazardous debris during bridge maintenance painting operations.

Research on environmentally compliant materials and testing of low volatile-organic-compound (VOC) coating systems focused on the relative cost/benefit of the durability of various paint systems based on performance data and relative costs of material and application. Materials testing projects—such as those on Performance of Alternative Materials in the Environment, Comparison of Laboratory Testing Methods for Bridge Coatings, and Environmentally Acceptable Materials for the Corrosion Protection of Steel Bridges—have had a direct influence on regulatory development and the development of measures for compliance by bridge owners.

These projects provided critical, long-term performance data for many new, environmentally compliant bridge paint materials and provided justification for bridge owners to move away from technologically old, lead-containing paints to new, more durable formulations that contain little or no toxic pigments and significantly less solvent. A project on Issues Impacting Bridge Painting, performed an extensive life-cycle versus initial-cost analysis for various bridge painting scenarios and presented results in a spreadsheet programme, to facilitate comparison of the various maintenance painting options based on a life-cycle cost analysis.

The projects Performance of Alternative Coatings in the Environment (PACE), Environmentally Acceptable Materials for the Corrosion Protection of Steel Bridges, Effects of Surface Contaminants on Coating Life, Maintenance Painting of Steel Bridges, Issues Impacting Bridge Painting, Methodology for Evaluation of Corrosion Control Coatings, and Comparison of Laboratory Testing Methods for Bridge Coatings have had aspects that addressed the need for higher quality and shorter term coating-durability data.

Common findings of these programmes with respect to cost considerations were:

- The relative cost of paint material is almost always insignificant when viewed in terms of the overall cost of the bridge maintenance job; and
- The advantage in the relative durability of the better coating systems often far outweighs the nominally increased cost of these materials at the time of application. In general, for moderately to severely corrosive environments, the most durable options in the coating material and in the surface preparation system will be the optimum choices from a life-cycle cost standpoint.

GUIDANCE AND SPECIFICATIONS FOR BRIDGE PAINTING, COATINGS, AND REMOVAL

Some states routinely involve state and federal environmental regulators in bridge painting design to help them satisfy the regulations. This has reduced compliance problems, cost increases associated with contractor force account work, and time delays; however, environmental regulators look for transportation agencies to come to them with good bridge painting plans and specifications. They are often reluctant to suggest specific methods and techniques, because it could compromise their regulatory role in the event of pollution problems.

The FHWA allows searches of specific state highway agency specifications using key words such as paint, coatings, and recycling. The site has a discussion feature to facilitate questions. Also, the FHWA users guide and spreadsheet *Cost Effective Alternate Methods for Steel Bridge Paint System Maintenance-Cost Model Users Guide* (FHWA Contract No. DTFH61-97-C-00026) can be used to evaluate cost effective bridge painting approaches. DOTs often use multiple factors such as adhesion test results, structure life, maintenance needs, layers of coatings, costs, and available funding to make bridge painting programme decisions.

The FHWA Publication No. FHWA-RD-94-100, *Lead-Containing Paint Removal, Containment, and Disposal,* provides information on the environmental and health regulations affecting the removal of lead-containing paints from steel bridges and includes a guide for waste reduction, control and disposal of the hazardous material generated by bridge paint removal operations. NCHRP Synthesis 251: Lead-based Paint Removal for Steel Highway Bridges and NCHRP 257: Maintenance Issues and Alternate Corrosion Protection Methods for Exposed Bridge Steel are earlier publications that remain valuable resources. Society for Protective Coatings (SSPC) Guide 6 (Containment design) and SSPC Guide 7 (Environmental monitoring) contain state-of-the-art guidance.

Missouri DOT Lead Paint Recycling

The Missouri Department of Transportation (MoDOT) maintains 7,138 bridges throughout the state of Missouri. Many of those bridges require lead paint removal and disposal prior to repainting. To safely accomplish this, the department contains and recycles the lead-paint waste through a lead-acid battery recycler. The lead is recovered in the smelting process and the steel grit or silica sand used as blast material and steel drums provide a substitute for fluxing agents used in the process. This has resulted in the complete recycling of hundreds of tons of former waste and saved valuable landfill space.

In 1996, MoDOT purchased two abrasive recycler systems. The abrasive recycler removed the lead paint from the steel grit by vacuum washes and magnetic separators. This reduced the amount of blast residue about 80 per cent. In recent years, water blasting of bridge paint has greatly expanded and further reduced the amount of blast material sent to the smelters by 99 per cent. In water-blasting, the water is recycled through a filter that removes the paint chips and the water is reused and at the end of the project, safely discharged to a permitted wastewater treatment facility. The paint sludge is recycled at a smelter.

MoDOT recommends the following practices when selecting a recycling company:

- Always check operating permits and qualifications of a prospective recycler.
- Check their history and ask a lot of questions about past problems.
- Review the environmental agency's files for any problems.
- If the files are sealed, find out why, and if you do not get a satisfactory response, reconsider that company as a prospect to be your recycler.
- Get a list of present and previous customers and talk to them about the company.
- Most importantly, visit their facility and get a tour of their operation.

METALLIZING

As a result of life cycle cost analyses, a few states have begun to explore metallizing as a coating option. Metallizing is a term used to describe thermal sprayed metal coatings. For corrosion control coatings on steel structures, metallizing refers to the thermal spraying of zinc or aluminum alloys as a coating directly onto steel surfaces. The coatings are created by using a heat source (either flame or electric-arc) to melt the metal which is supplied as a wire or in powder form. An airstream sprays the molten metal onto the steel surface in a thin film. Once the metal strikes the steel it resolidifies to become a solid coating. Metallized coatings provide corrosion protection to steel by sacrificial and barrier protection.

The coating itself provides a barrier between the environment and the steel surface, especially when applied in combination with conventional sealers as topcoats. Due to the electrochemical reaction between steel and zinc or aluminum in an aqueous and salt-contaminated environment, these coatings tend to "sacrifice" themselves to protect the steel at the site of any damage in the coating.

This sacrificial protection is similar to the protection provided by zinc-rich primers or galvanizing. According to the Turner Fairbank Research Center, metallizing has been reported to be highly effective in numerous research projects and in observation of historical applications.

Many reports have proclaimed that a metallized structure will last 25 to 40 years with no need for maintenance touch-up. This greater life expectancy and higher effectiveness brings a higher initial cost, but a potentially lower long term cost. A project recently completed by the Illinois Department of Transportation (IDOT) incorporated metallizing technology as an experimental feature. For this project, the structural steel was metallized in the shop then transported and erected in the field.

There have been several bridge rehabilitation projects around the country which have had metallizing done in the field on existing steel, but only a few have been completely metallized in the shop. The production rates, cost concerns, logistics, handling and construction issues were examined in this study in order to assist in determining the feasibility of and issues involved in using metallizing in shop applications. The project was considered successful and equipment advances enabled metallizing to be done much faster than in the past. The project illustrated that under life-cycle costing, metallizing can be advantageous despite its cost, on this project, of $9.18 per square foot versus $2.00-$2.50 per square foot for painting.

Overcoating

As an alternative to removal, some toxic based paint is in a condition that permits an overcoating of paint to effectively contain the toxic material and protect the steel. Critical variables which determine the success or failure of an overcoating job include: the condition of the existing paint, the extent of corrosion on the substrate, the level of surface cleanliness achieved, and the environment of exposure.

A recently completed FHWA-sponsored study of various overcoating materials applied to bridge structures in various parts of the country resulted in the following general results:

- Multicoat systems, overall, performed better than single coat systems, with three coat systems showing generally better performance than two coat systems. This result is thought to be directly related to the occurrence of pinholes during brush application of maintenance coating materials; however, it shows a measurable benefit of the application of multiple coats in a realistic scenario. Other studies have shown similar results.
- Coating materials performing well in the study were three-coat moisture-cured urethane systems, and three-coat epoxy based systems using a penetrating low-viscosity sealer as a primer. In addition, two separate three-coat low-VOC alkyd systems performed well. Of the single coat systems, the coatings based on calcium-sulfonate alkyd resins did best. Similar generic results have been found by other investigators.
- In general, coatings that did well at any one test site did well at all four, diverse test sites 'indicating some measure of surface tolerance, or "overcoating acceptability" for specific paint materials. Those materials that failed badly and early at any one site generally failed at more than one site. This would lend

support to the use of patch tests as screening for acceptability of a particular overcoating material on a particular structure.

- In the subject testing, failures were of two varieties, (1) early coating disbondment due to incompatibility of the overcoating material with the existing paint, and (2) rust through of the newly applied overcoating material at areas where the coating is over bare steel or existing rust.

The performance of a newly applied overcoating is highly dependent upon the condition of the existing coating over which it is applied. Overcoating over existing aged paints that are often brittle and loosely adherent can pose risks of early failure and large scale disbondment. The applied overcoat applies added stress by adding physical weight to the existing coating, shrinkage during drying, different expansion and contraction rates between two different paint systems with ambient temperature cycles, and softening of the existing coating due to solvents in the overcoat.

The following steps can minimize this risk:

- Prior to deciding to overcoat a particular structure, assess the condition of the existing steel and paint system. Condition can include extent and distribution of rusting, and adhesion. The extent of metal loss due to corrosion and the extent and distribution of paint breakdown are important in determining the scope of the work required to clean and overcoat the structure. Paint failure confined to specific definable areas is easier to clean and overcoat than general paint failure over the entire structure. Adhesion should be assessed using either standard test methods (ASTM D4541 or D3359) or by attempting to cut and lift the coating with a knife blade. If the coating lifts easily or crumbles at the tip of the knife blade, then application of overcoating paints should be considered high risk.
- Conduct a representative patch test of the new material over the existing material using representative

maintenance painting practices. Allow this patch to weather several seasons and assess the compatibility of the systems. Research has shown that reasonably good performance is achievable with various different overcoating materials. A patch test will help eliminate coatings grossly incompatible with the existing paint on the structure.

- Where possible, consider using coatings similar to those currently on the bridge. Some investigators have seen good results from the application of newer, but generically similar alkyd coatings over older existing alkyd coatings. Some of the older technology coatings may present a compromise in overall durability when compared to newer coating materials, but in a maintenance mode, the compatibility of these materials is a definite advantage.

After the existing paint has been sampled and a plan of action developed for an appropriate paint system, spot painting can be performed by maintenance staff using the following guidelines:

- If the paint has not been proven to be lead/chromium/cadmium free, treat it as if it were hazardous and brief personnel accordingly (refer to OSHA pamphlet 3126).
- Personnel removing the paint should wear coveralls, gloves, goggles, and a certified, properly fitting respirator for protection.
- Ensure a containment system is arranged to catch and retain all the paint removed.
- Apply an approved chemical paint remover on the desired area and allow it to stand.
- Remove the paint with hand tools or scrapers that do no not cause the paint particles to become airborne.
- Dispose of hazardous paint at the nearest district headquarters yard at the hazardous waste site in the area marked for toxic paint. Inform district personnel.
- Prepare and clean the now paint-free surface and repaint the area with an approved paint.

- Ensure that personnel do not eat, drink or smoke until they are finished and have washed and properly disposed of all contaminated tools and clothing.
- Cracks in any bridge component should be evaluated and corrective action taken. A crack in any steel component must be promptly reported to the Bridge Engineer and corrected per recommendation. Cracks in wood or concrete should be evaluated first and then cleaned and sealed with an appropriate crack sealant. Larger cracks in concrete should be sealed with polymer or epoxy based sealants. Small shrinkage cracks and all concrete surfaces where the concrete is not a specialized or high density concrete such as latex modified or silica fume concrete should be treated with a silane based sealant.

Painting operations generate dust, solvent fumes, and noise. Every effort should be made to minimize the impact of these operations on the surrounding community.

Inspecting the Structure and Preparing Equipment

Inspect the structure paying particular attention to areas of localized rust because these are the areas that have shown to be prone to premature coating failure. Extra effort should be made to ensure that both the proper degree of surface preparation and the proper coating thickness are achieved in these areas. The presence of mill scale under the existing paint indicates a potential need for additional surface preparation. If mill scale is observed and abrasive blasting is not specified, the project Engineer should be notified since abrasive blasting may be required.

- Inventory, inspect, and calibrate the equipment.

Surface Preparation

Proper surface preparation and proper paint application are the two most important factors needed in a high quality job that will avoid peeling paint and future environmental contamination. It has been estimated that 75 per cent to 80 per cent of all premature coating failures are caused, partially or completely, by deficient surface preparation and/or coating application. Cleanliness is essential since the presence of oil, grease, dust, or soil prevent the paint from bonding. Mill scale, rust, and the existing paint may increase the chance of failure of the new coating.

Clean surfaces must have an appropriate anchor pattern (surface roughness). This roughness helps the new paint to mechanically bond to the surface, promoting adhesion. The Society for Protective Coatings (SSPC) has developed a nomenclature for the different types of surface preparation methods. Best practices for each are noted in a bulleted list following description of the SP level practice. FHWA has two studies that speak to this issue.

Hand Cleaning

SP-1: SP-1 denotes "solvent" cleaning and can refer to solvent wiping, water washing, or steam cleaning. The surface is cleaned to remove oil, grease, etc. This must be done prior to ALL other cleaning operations as some final surface preparation methods will actually force the contaminants into the steel, which can lead to poor bonding and premature failure.

- Clean, lint-free rags and clean solvent should be used to avoid the spreading of contaminants.
- Once the contaminants have been visibly removed, a final wiping should be done with clean rags and solvent.
- Workers should wear goggles, protective clothing, rubber gloves, and petroleum jelly on exposed body parts and should be equipped with appropriate respirators to avoid hazardous fumes.
- Benzene and carbon tetrachloride are poisonous and should not be used as solvents and neither should materials with low flash points such as gasoline, methyl-ethyl ketone (MEK), and acetone. Consult the Materials Safety Data Sheet to determine the specific hazards and protection procedures to be followed for the solvent being used.

SP-2: SP-2 denotes hand tool cleaning. Hand tools are used to remove loose mill scale, loose rust, loose or otherwise defective paint, weld flux, slag, and spatter. This is done by brushing, sanding, chipping, or scraping the surface. Tools used include wire, fiber, or bristle brushes, sandpaper, steel wool, hand scrapers, chisels, or chipping hammers. Tightly adhering rust, mill scale, and paint are allowed to remain. This method is generally confined to small areas.

- Verify that the level of cleanliness noted in SP-1 has been achieved. Pay particular attention to the problem areas such as the top side of bottom flanges, the backside of nuts and bolts, the interior of box beams, and those areas where climbing is difficult and access is limited.

SP-3: SP-3 denotes power tool cleaning. This is very similar to SP-2 except that power tools are used instead, thus making this a more viable and efficient cleaning method for larger areas.

- Check that the power tools have not placed any oil or grease back onto the surface. If they did, the surface should be re-cleaned per SP-1.
- Paint, rust, or millscale that can be removed with a hand scraper should not remain after a proper SP-3 surface preparation. A "dull putty knife" can be used to assess the acceptability of the surface.

SP-11 denotes power tool cleaning to bare metal. This method uses power tools to remove ALL paint, rust, and millscale and to roughen the surface to promote paint adhesion. SP-11 offers performance advantages over SP-2 and SP-3, which result in an irregular surface of bare steel, rusted steel, mill scale, and paint but it tends to be quite expensive because of the labour involved.

Blast Cleaning

Blast cleaning is the most effective method for surface preparation is blast cleaning. Blast cleaning is broken down into four levels according to the desired condition of the base metal. Blast cleaning does not get rid of oil and grease, which is done by solvent cleaning. There are many types of abrasive on the market. Recyclable abrasives have gained popularity in recent years because their use can reduce waste handling and disposal by 90 per cent.

Blasting Using Recyclable Abrasive

In this system the abrasive is accumulated after usage, cleaned, and reused more than one time. Recyclable abrasives must be hard and durable. Thus metallic material is typically used, and typically requires special equipment to collect, classify, separate, and convey collected waste residue. Also, since the abrasive is harder, contractors must pay close attention to abrasive gradation to keep a cleaned surface profile within acceptable ranges. A contractor must closely monitor the separation process. It is important to completely remove all fine material from abrasives. If the abrasive is improperly or incompletely cleaned, dust concentrations within the containment can be adversely affected. Several methods are available in the industry to filter discharged air from the system. Often systems using water for blasting or water filters to remove particulates are not acceptable as the water then becomes another different waste for disposal; however, MDSHA uses a pressurized filter system to remove particles from bridge cleaning waste water.

The Golden Gate Bridge Authority in the San Francisco Bay area uses filters to remove particles from bridge painting waste water. As with all open blasting operations, the recycled abrasive method must be fully contained. Costs associated with recyclable abrasive include additional equipment and increased initial abrasive costs. This is offset by increased cleaned surface area per unit of abrasive (some times up to 100 cycles) and reduced volume of waste produced.

The Missouri DOT has developed a model recycling programme for abrasive lead blast. MoDOT no longer does abrasive blasting with Department forces; the agency has tried to create incentives for contractors to develop cost saving techniques and technologies for recycling its abrasives. All

MoDOT bridge painting waste containing lead collected by Department crews is taken to lead smelter for recycling.

Closed Abrasive (Vacuum) Blasting

Closed abrasive blasting or vacuum blasting allows dust, abrasive, and paint debris to be vacuumed simultaneously with the blasting operation. Debris is separated for disposal and the abrasive is returned for reuse. Typically, hard metallic abrasives are used for this system. Vacuum blasting equipment is expensive; however, both worker exposure to dust and environmental emissions can be minimized if operations are conducted properly.

Special Provisions may allow vacuum blasting to be conducted without requiring full containment. Once again, systems that uses water or water filters cannot be used. Vacuum blasting is limited by its reduced production rate and operational problems cleaning edges and irregular surfaces. To be completely effective, the whole nozzle assembly must be sealed against a surface to maintain proper suction for the vacuum operation.

Fig. 2.6 Shrouded Power Tools

Shrouded power tool technology has proven an effective engineering control that effectively prepares structural steel for new coatings, while simultaneously controlling all emissions in excess of 99.5 per cent. On the Woodrow Wilson Bridge outside of the District of Columbia, this eliminated the need for containment and respirators and reduced the potential for lead poisoning to the environment or workers in the first place.

These DOE, EPA, OSHA, and HUD-tested and approved tools utilize a mechanical, air-driven process that cleans surfaces to a bare substrate, while a High Efficiency Particulate Air (HEPA) filtered vacuum collection unit, the VAC-PAC, simultaneously captures dust and debris and transports it into an on-board 55-gallon drum. The Pentek system offers a fully integrated deleading system that removes, collects, drums, and seals the waste in a single step process for safe disposal.

The shrouded power tools allow the workers to prepare the structural steel to a bare metal finish comparable to a Steel Structures Painting Council SP 6 specification, Commercial Blast Cleaning. Pneumatic-powered rotary scalers and needle guns are being operated simultaneously—the former for the rapid deleading of large, flat surfaces and the latter with adjustable shrouds and pivoting head for access to hard-to-reach areas, such as around bolts and angles, or in corners.

Fifty-foot vacuum hoses attached to the tools convey all removed dust and debris down to the HEPA-filtered vacuum and waste collection unit, which is stationed on the deck of a barge below. The 100 per cent mechanical coatings removal system minimizes the amount of waste for disposal, as well as the degree of the owner's liability and disposal costs, by adding nothing to the waste stream and collecting only the dust and debris of the coating itself.

A typical lead abatement project employing Pentek's system deposits 2,500 square feet of surface in a single waste drum. An independent company conducted air quality monitoring for the project with personal sampling and high volume environmental monitors, which were placed strategically at the site. Two separate air readings with the high volume environmental monitors were conducted six weeks apart, one on land and one on the barge. The efficiency of the dustless power tool system was verified by air sampling results of two to three micrograms per cubic meter, far below both OSHA's Permissible Exposure Limit of 50 micrograms per cubic meter and the action level of 30 micrograms per cubic meter per eight-hour exposure limits.

SP Levels of Blast Cleaning

The following methods are presented in order of ascending cleanliness (SP-5 is most clean):

SP-5 - SP-5 denotes white metal blast cleaning. This level of cleaning is costly and is rarely specified for use on bridges:

- The resulting surface should be free of oil, grease, dirt, rust, mill scale, all paint, and foreign matter leaving only a uniform grey-white colour.

SP-6 - SP-6 denotes commercial blast cleaning:

- The resulting surface should be free of oil, grease, dirt, all rust, mill scale, paint, and foreign matter (except for slight shadows) streaks, or discolorations caused by rust stains, mill scale stains, and tight residue of previous coatings.
- At least two-thirds of each 150-cm² (9 in²) area must be free of all visible residue and the remainder limited to those discolorations just mentioned.

SP-7 - SP-7 denotes brush off blast cleaning:

- The resulting surface should be free of oil, grease, dirt, loose mill scale, loose rust, and loose coatings, retaining only tightly bonded mill scale, sound rust, and previous coatings.

SP-10 - SP-10 denotes near-white blast cleaning:

- The resulting surface should be free of oil, grease, dirt, rust, mill scale, paint, and any foreign matter (leaving only slight stains from rust and mill scale).
- At least 95 per cent of each 150-cm² '9 in², area should be free of all visible residue with the remainder limited to slight discoloration.

SP-12–SP-12 is the standard for pressurized water blasting: The most common surface preparation specified for bridge use is SP-6, which requires SP-1. When repainting an existing structure, the Specifications may call for SP-2 or SP-3 in areas of limited accessibility. Recent research indicates SP-10 may be more cost effective than SP-6, particularly in more corrosive environments. Inspection should verify that both the proper level of cleanliness and the proper anchor pattern have been achieved. Blasting operations require that several additional checks occur. The contractor's equipment and material must be checked along with the resulting anchor pattern.

Inspection of the Abrasive

The contractor will select the abrasive to be used based on the specified anchor pattern. The chosen abrasive should be free of toxic heavy metals such as lead, chromium, and cadmium and should not contain any free silica (sand) either. A sieve analysis from the abrasive supplier should be requested prior to delivery of the first load of abrasive.

Once the abrasive is on site, obtain a sample of the stored abrasive material. It should be stored in a dry environment and should be clean, uniform, and free of any sign of moisture. To check, drop some of it into deionized water and shake. Watch for a film of grease or oil indicating the presence of contaminants. Keep a small sample of abrasive from each subsequent delivery. This will allow for a future analysis in the event that changes occur in the anchor pattern.

Inspection of the Air Supply

This is necessary to ensure that the air supply is not introducing neither contaminants that will be embedded in the steel nor oil or water into the system. Inspect the air compressor for contaminants. The compressor should have moisture and oil traps on all lines. Shut off the flow of abrasive. Place a white blotter cloth in the air flow. It should be placed approximately 0.6m '24 inches, from an outlet downstream from the oil separator and moisture traps. Let free air flow for two minutes. Check for visible contaminants in the air flow; if there are any, corrective action is needed. This test should be repeated every four hours or more frequently when the humidity is high.

Blasting Pressure

The blasting pressure should be at least 620 kPa (90 psi); any less than this can result in a lower anchor pattern and in slower production. However, jobs that use recyclable steel grit often use higher pressures. All high pressure air supplies and devices should be gauged for easy reading. For blasting, the critical pressure is located at the end of the blast nozzle. This pressure will be lower than that measured at the air supply due to loss in the hose. Hence, limiting the length of air hose is often a critical factor in the efficiency of a blasting operation. A pressure needle gage may be used at the nozzle to measure the true blast pressure.

Inspection of SP-6 and SP-10

Once again, verify that the level of cleanliness noted above has been achieved including the referenced problem areas. The individual DOT Specification may provide a set of visual standards from either the National Association of Corrosion Engineers (NACE) or the Society for Protective Coatings (SSPC) to aid in this effort. Interpretation of the visual standards can take some discretion as well as some practice. Both SP-6 and SP-10 standards require the removal of ALL paint, rust, and mill scale. The only difference is in the amount of staining allowable on the bare steel surface.

Inspection of the Anchor Pattern

The anchor pattern needs to be checked to ensure that proper paint adhesion will occur. Profile inspection requires the use of a micrometer and replica impression tape. Comparison coupons can be used for a qualitative visual comparison of the profile.

Monitoring

Air and soil monitoring are becoming more common on lead removal jobs. Monitoring protocols differ, and there is no current consensus on what should be done in this area. FHWA's study tour on Bridge Maintenance Coatings Environmental and Worker Protection Practices reportedly uncovered no requirement or specification for environmental air monitoring for sources other than stationary sources, and bridges are not considered stationary sources; environmental air monitoring requirements for abrasive blasting of lead-containing paint from bridges and other structures were not encountered in the United States or Europe, despite increasing ambient monitoring for total lead in dust and particulate size (PM10) in this country.

Nevertheless, many state DOTs, as well as numerous city and local authorities require environmental air quality monitoring for abrasive blasting operations as part of their specifications and policies. FHWA's study notes that current FHWA research could lead to "more reasonable and applicable protocols for environmental monitoring during bridge-painting operations."

FHWA's study found that soil lead level monitoring before, during, and after the project is usually required. The Swiss have noted that as much as 70 micrograms of lead per m^2 per day may be deposited during movement and teardown of containment systems. Allowable levels of total lead in soils were found to be as low as 50 ppm. Some states are requiring pre- and post-job soil monitoring for lead contamination, but the requirement is not universal.

Characterizing the contamination level of soils surrounding bridge job sites and especially the specific source of lead in any one location is difficult at best. Field monitoring and research is currently underway to attempt to better define appropriate soil-sampling protocols.

- Soil sampling near and under containments is generally a good idea from a liability standpoint. Check for signs of surrounding ground or water contamination.
- Air monitoring becomes more important if sensitive public access is nearby.

PAINT SELECTION, STORAGE, HANDLING, AND MIXING

- Use paints with maximum useful lifetimes, where toxicity is acceptable, not lowest cost, to maximize the time between repainting.

- Verify that the paint has not exceeded its shelf life. Shelf life is the length of time, from date of manufacture, that a paint will remain usable when stored in its can. Consequences of exceeding the shelf life include: gelling, odour, changes in viscosity, formation of lumps, pigment settling, and colour and liquid separation.
  - Check the date printed on the can with the shelf life to make sure the paint has not "expired." Some suppliers use a special code on the can which contains the date of manufacture. It may be necessary to call the supplier to read the code and assure that paint is fresh. Two-component paint systems often have a different shelf life for each component.
  - If the contractor desires to use this material that has exceeded its shelf life, he/she may submit a sample to the manufacturer's laboratory for analysis and possible re-certification. The contractor should not be allowed to use the material in question until written certification is received from the manufacturer.
- Follow good storage practices and verify that the paint is not stored in areas subject to temperatures beyond the recommended limits. Going beyond the acceptable temperature range can cause changes in viscosity and shelf life. Water-based paint will spoil when stored below freezing. Solvent-based paint, on the other hand, may gel or become flammable or explosive when stored at high temperatures.
  - The contractor's storage site should be monitored with a high/low thermometer. Contractors often like to store the paint on site in a trailer. This is generally not a good idea because these trailers tend to get very hot during the summer and have limited ventilation. Paint should be stored in a climate-controlled environment.

– Each lot of paint should be stored together.
– Two-component systems should be stored close to each other, but be distinguishable from one another.
– If the paint will be stored over several months, the cans should be inverted at monthly intervals to avoid excessive settlement and ease future mixing.
– When opening the paint, the oldest paint should be used first. Look for signs of aging listed under shelf life.
– Note the required temperature range for proper storage. Adherence to the temperature requirements noted on the Product Data Sheet is essential.

Verify that Pot life has not been Exceeded

Pot life refers to the length of time a paint is useful after its original package has been opened or, for two-component systems, the length of time after it has been mixed. Pot life is temperature dependent. The pot life on the Product Data Sheet is generally for 21°C 70°F.

Contact the manufacturer for additional pot life information if the paint has been stored in temperatures outside of this general range. Exceeding the pot life can result in sagging of the fresh paint along with poor performance attributable to film porosity and/or poor paint adhesion. Two-component paints tend to become unworkable at or beyond their pot life.

Ensure Proper Mixing and use of Thinner

Different paints have different mixing requirements. The instructions on the Product Data Sheet should be strictly followed. Thinner is a liquid added to the paint at the time of application to modify its viscosity. The Product Data Sheet will indicate the specific type and maximum amount of thinner to be used.

- Upon opening the can, check the surface of the paint for "skinning over" of the paint. Any skin should be removed prior to mixing.

- All paint must be thoroughly mixed in a clean container.
- Check the bottom of the original can for evidence of unmixed pigment.
- For two-component paints, verify that they are mixed in the proper proportion. The mixing operation should be witnessed and documented.
- Unused paint that will be used the next day should not be left in buckets or spray pots. It should be placed in a container and re-mixed prior to use.
- Thinner should be used only to achieve optimum viscosity for proper application and is not always necessary. Do not exceed recommended maximum use.
- Witness and document each and any addition of thinner. Adding too much thinner can prevent proper application thickness and cure of the paint and may result in the mixture exceeding acceptable limits for volatile organic compounds (VOCs).

Verify Drying and Curing Times

Drying time refers to the length of time a coating is sensitive to local damage. Curing time refers to the length of time it takes for a paint to reach structural integrity and be ready for service. The drying schedule on the Product Data Sheet will show how long it takes until the paint is dry to the touch, dry to tack free, and dry to recoat. Dry to the touch implies the paint won't collect dust; tack-free implies the paint does not feel sticky and can be handled without damage; dry to recoat implies the time needed to dry until the next coat of paint can be applied.

Drying times vary significantly with temperature. This is particularly important in determining when the next coat of paint can be applied. Recoating before enough time has passed can seriously affect the curing and integrity of the layer being overcoated. Some paints, particularly two-component paints, have a maximum time to re-coat as well. Exceeding this could jeopardize the adherence of the top coat.

- After painting, inform the contractor of the estimated

time that should be allowed for the paint to cure. Do not allow another coat to be put on until the appropriate amount of time has elapsed per the existing weather conditions.

- Carry out storing, mixing and cleaning operations on land.
- Transport paint and materials to and from job sites in containers with secure lids and tied down to the transport vehicle.
- Do not transfer or load paint near storm drain inlets or watercourses.
- Test and inspect spray equipment prior to starting to paint. Tighten all hoses and connections and do not overfill paint container.
- Plug nearby storm drain inlets prior to starting painting where there is significant risk of a spill reaching storm drains. Remove plugs when job is completed.
- Cover nearby storm drain inlets prior to starting work if sand blasting is used to remove paint.
- Perform work on a maintenance traveller or platform, or use suspended netting or tarps to capture paint, rust, paint removing agents, or other materials, to prevent discharge of materials to surface waters if the bridge crosses a watercourse. If sanding, use a sander with a vacuum filter bag.
- Capture all clean-up water, and dispose of properly.
- Recycle paint when possible. Paint may be used for graffiti removal activities. Dispose of unused paint at an appropriate household hazardous waste facility.
- Keep all materials securely locked up, to avoid vandalism and accidental spills into the watercourse.
- Hold a pre-painting meeting with the contractor, addressing the following issues. Minutes should be kept and a copy should be given to all meeting participants, documenting understandings and any agreements reached.

- The nature of the work and its effects on the surroundings, including possible mitigation measures.
- Contractor's method of operation, including equipment and personnel.
- Contractor's schedule. Discuss weather-related concerns.
- Contractor's job-specific worker health and safety plan (if lead paint is present).
- Proper storage of material and equipment.
- Location of recycling and dust collection and storage equipment.
- Inspector safety, including provision of safe access and safety from lead contamination.
- Inspection and measurement procedures, including control points.
- Identification and treatment of inaccessible areas.
- Product Data Sheets and Materials Safety Data Sheets for all relevant materials.
- Visual standards to be met. Discuss contractor's preparation of field reference sections.

Containment and Use of Enclosures

Lead was a common component of industrial paints until the 1980s, and many of the steel bridges in the highway system are still coated with paint that contains up to 50 per cent lead by weight. High lead-containing primers can often be identified by their red or bright orange colour. However, not all red and orange paints contain lead, and some paints of different colours can contain a significant amount of lead. Lead is hazardous to humans if it is inhaled or ingested and a relatively small amount of ingested or inhaled lead dust can elevate a person's blood lead level.

Proper respiratory protection should be worn to protect against lead hazards. "Proper" protection consists of either air-fed, positive pressure respirator hoods (as worn by abrasive blasters), or negative pressure, filter-cartridge respirators. Filters should be colour-coded bright pink for fine dust particulate (HEPA) filters. The required level of respiratory protection depends on the concentration of lead in the breathing air, and on the amount of time the worker is exposed.

For most short-term inspections of jobsites without ongoing blasting, or outside of containments, a half-mask with appropriate HEPA cartridge is enough. However, while inside of containments during or immediately after abrasive blasting, an air supplied hood is likely to be required. A containment system, or enclosure, is needed to prevent both lead and other debris generated during surface preparation activities from entering the environment and to facilitate its gathering and disposal.

Enclosures are generally made up of combinations of cover panels, scaffolds, supports, screens, and tarps. The complexity of any given enclosure will vary depending on the method of paint removal being employed and the degree of surface preparation that is specified. For a simple scraping operation ground-covering tarps may be sufficient while for a blasting operation, the enclosure could be a designed structure with a negative pressure ventilation system.

Containments for abrasive blast (and other paint removal) operations are designed to protect the surrounding environment and the public from debris (flying abrasive) and potentially hazardous material (lead-containing dust) during a paint removal operation. In addition, these containment structures are intended to help contain and collect the lead-containing debris for proper treatment and disposal.

As the use of containments for paint removal jobs has become more common over the past several years, the design of containment structures has evolved. Currently, there are standard features of each containment method, but in large part, containments are custom-designed for each bridge job. The standard features are described in detail in the "SSPC Guide 6 - Containment." This guide is not all inclusive, but it is the industry standard for description and classification of paint removal containments. In addition, several States have their own classification systems, but most of them are somewhat similar to those in the SSPC Guide 6.

It is important to remember that the purpose of a containment is to do a conscientious, "state-of-the-practice" job in containing and collecting debris. With appropriate specifications, and designs, and a cooperative effort between the owner and contractor, very near 100 per cent containment and collection of debris can be approached. Containment of abrasive blast jobs involving lead are mandatory because the law requires collection of the hazardous waste. While there is no specific rule governing the "fugitive emissions" of lead-containing dust, this dust can be controlled by designing and maintaining the containment and ventilation system properly.

Bridge maintenance involves the installation of safety nets, tarps, enclosures, barges or other means to catch paint chips and removed debris, water and abrasive blasting to remove old paint, rust, grease etc., application of rust inhibitors, primer paint (zinc aluminum or lead), mid-coat paint (epoxy) vinyl or lead, and final coat (epoxy) polyurethane, vinyl or lead. All of these activities and products are detrimental to aquatic life in the stream below and need to be prevented from reaching the stream.

Components of a Containment System

Containment System Structure

Containments can be scaffolded from the ground or rigged to hang from the bridge structure. The key issues to consider are structural integrity under wind load, abrasive waster load, and dynamic loads on the bridge. Access, air movement, and visibility should also be considered.

- Use custom built enclosures to confine and capture the abrasives, old paint chips and paint where possible.
- Erect shrouds around working areas and suspending nets and tarps below bridges to catch debris from abrasive removal of old paint and over-spray from painting, where wind conditions permit. The work

area should be clearly distinguishable from the surroundings.

- Anchor tarps to barges below and enclosing the bridge above to confine debris, where the bridge deck is not too far above water level. Using barges and booms to capture fugitive floating paint chips and debris netting is not adequate.
- Tarps should be overlapped with seams fastened and should be in good condition and free of holes.
- During blasting operations with negative pressure, the tarps should have a concave inward appearance. They should never appear to bulge during blasting.
- The containment should be tightly sealed to prevent any dust from escaping. Continually evaluate and/or perform field checks on the effectiveness of any containment. Watch for signs of dust escaping the containment and/or dust being discharged from exhaust system. Check the ground around the containment.
- The containment must also be able to support workers, construction loads, spent abrasive loads and wind load without placing undue stress on the bridge.
- The containment should be constructed in accordance with the approved plan.
- Use vacuum or suction shrouds on blast heads to capture grit and old paint where possible.

Ventilation

For work inside an enclosure air movement is necessary to avoid a build-up of dust. High dust concentrations impair visibility and increase hazardous exposure levels to workers. Without ventilation, workers and inspectors will not be able to see within minutes of blasting commencing. Ventilation also reduces the concentration of lead-dust in the work environment and makes clean up operations prior to painting easier.

Check for air movement with an anemometer, or get a rough idea of the amount of airflow by using a smoke bomb.

Air movement is dependent upon the capacity of dust collectors, the volume of air input by makeup fans and blast nozzles, and interferences to airflow caused by the bridge structure itself. Ventilation ducting efficiency depends on the duct diameter and length, and on minimizing the number of sharp bends in the duct.

Location of Dust Collectors

Airborne emissions are often highest adjacent to dust collectors. While emissions should be minimized, they are unavoidable.

- Dust collectors should be located in areas where emissions will have minimal effect on sensitive surrounding environmental or public areas.
- Dust collectors should be operated at the rated capacity or at a capacity consistent with the ventilation design of the containment system.

Lighting

Proper lighting is often neglected. Inadequate lighting poses obvious safety concerns as it makes proper surface preparation and painting almost impossible. The potential for escaped lead-contaminated air emissions during blasting operations also warrants the usage by abatement workers of full facepiece respirators operating in positive-pressure mode.

Shrouded power tool technology has proven an effective engineering control that effectively prepares structural steel for new coatings, while simultaneously controlling all emissions in excess of 99.5 per cent. In other words, the need for containment and respirators is eliminated or drastically reduced by preventing the potential for lead poisoning to the environment or workers in the first place.

PAINT APPLICATION AND SPRAYING PRACTICES

Once the proper level of surface preparation has been achieved and the quality of the coating system has been verified, the contractor is ready to paint. To prevent "rust-back" of the cleaned surface, the first coat of paint (primer) should be applied as soon as possible (within a few hours) after blast cleaning. Painting should begin at a practical time to avoid weather changes that could cause significant changes in the surface condition of the steel, nightfall. Painting just before the onset of poor weather is not advisable. The current and expected weather, along with the curing time of the paint being used, should be considered prior to beginning the application process.

To ensure that the paint is applied and allowed to dry and cure under reasonable environmental conditions, the following environmental conditions should be followed every four hours:

- *Temperature*: The Specification will place limits on the ambient temperature to ensure proper curing. Most Specifications will require the temperature to be between 4°C or 10°C and 38°C '40 or 50°F and 100°F.
- *Relative Humidity*: Again, due to of curing requirements, the Specification will limit the maximum permissible relative humidity, which is commonly limited to 85 per cent.
- *Dew Point*: Using the relative humidity, determine the dew point. The temperature of the steel should be at least 3°C '5°F, higher than the dew point. This "dew point spread" is used to ensure that no moisture is present on the steel prior to paint application.
- *Surface Temperature:* The surface temperature of the steel should not exceed 52°C (125°F) during the painting process, and, again, it should be at least 3°C (5°F) higher than the dew point.
- *Wind*: Heavy winds can cause problems. Airborne overspray, for example, may be carried onto adjacent houses, cars, etc. and can result in premature drying of the paint. If heavy winds are present, it may be best to delay the painting operation or to restrict spray application.

DEBRIS STORAGE

The disposal of equipment and materials used to remove

existing paint is also expensive. The waste must be placed in storage containers approved by the EPA. The contractor had two options for removing and handling the existing paint. Each option is designed to either make the removed material nonhazardous or reduce the amount of hazardous material generated; depending on the removal technology used, the waste is used as a recycled material or placed in a hazardous-waste landfill after being treated.

DOTs should and generally do require all waste generated during removal operations to be sampled and analysed by the contractor and submitted to a laboratory for Toxic Characteristic Leaching Procedure Testing (TCLP) for eight environmentally regulated heavy metals typically found in paint and abrasive wastes. Paint debris is classified as hazardous due to the characteristic of toxicity, if after testing by TCLP, the leachate contains any of the elements in the concentrations equal to or greater listed levels. Other elements, chemicals, and characteristics can also cause a material to be hazardous as defined in 40 CFR 261, so best practice requires that:

- No other waste be mixed with paint waste generated during the cleaning process.
- Accumulated wastes shall not be removed from the temporary storage area without proper documentation.
- For all projects involving the removal of paint wastes, some form of manifesting is required.
- Recycling (off-site) or proper disposal of hazardous abrasive-blast media and use of approved haulers to transfer the contained waste to authorized treatment and/or disposal facilities
- The production of hazardous paint-removal waste should be minimized by the use of recyclable abrasive and the waste generated should be treated by effective methods to ensure its stability in waste containment sites.

Lead-contaminated paint waste can be classified as hazardous material. As such, it is subject to strict disposal requirements. The contractor may wish to temporarily store barrels containing this waste on site prior to hauling them to an approved disposal site.

- If the barrels are stored on site, regulations restrict such storage to 90 days. Some entities, such as New York City Transit, voluntarily limit temporary storage of hazardous wastes to 45 days, instead of the allowable 90 days.
- Barrels should be clearly marked as containing hazardous waste.
- The barrels should be stored in a location inaccessible to the public, and they should also be in a location where they are not at risk of being hit by traffic.

QUALITY ASSURANCE AND PUBLIC OUTREACH PROGRAMMES

In 2001, Illinois' Department of Transportation (IDOT)revised its overall painting policy along with its environmental and containment specifications for all repainting maintenance. The most significant change in the specifications was related to quality control and quality assurance responsibilities. The Kentucky Transportation Cabinet has also been a national leader in the development of a quality assurance for bridge painting and maintenance.

As a proactive quality assurance measure, the Indiana Department of Transportation outline for bridge painting pre-bid meetings, *Specification and Pre-Bid Conference Content Review,* is used by the Indiana DOT environmental and safety professionals to make pre-bid conference presentations. The outline helps InDOT staff ensure that important environmental and safety topics are always presented at these conferences. Public outreach programmes have become a part of bridge painting and lead removal projects. The Port Authority of New York City provides a good example programme.

3

Maintenance Facilities Management

PLANNING AND PRIORITIZING

Facilities management encompasses a broad range of activities, including:

- Storage, repair, and maintenance of vehicles, equipment, and related support materials
- Fueling and washing of vehicles and equipment
- Maintenance of buildings, stormwater drainage systems and landscaping
- Storage of sand, salt, asphalt, rock, and pesticides
- Storage of wastes generated on site
- Bulk storage of sediment, litter and debris generated by road maintenance activities

Environmental stewardship in the course of these activities requires both structural and non-structural management practices. Examples of non-structural practices include procedures for performing operational activities, such as salt/sand mixing/loading that requires removal of all salt from the area surface after loading. The installation of a physical device that alters the release, transport, or discharge of pollutants from surface storm or melt water or facility-generated shop floor drain or washbay effluent is a structural practice. Many environmental stewardship practices at maintenance facilities have to do with protection of water quality. EPA regulations have long required facilities to obtain National Pollution Discharge Elimination System (NPDES) permits for discharges, especially washbay and shop floor drain effluent discharges to the waters of the State. Such permit obligations arise under the Industrial Permitting portion of NPDES, and have received increased attention as state regulatory agencies have expanded beyond their initial focus on manufacturing facilities in implementation of this programme.

Brief summaries of federal water quality and wetlands requirements applicable to the transportation community are available at Center for Environmental Excellence by AASHTO.

POLLUTION PREVENTION PLANS

Facility Pollution Prevention Plans (FPPP) are typically developed for each maintenance facility owned or operated by a DOT. The FPPPs describe the activities conducted at the facility and the management practices to be implemented to reduce the discharge of pollutants in stormwater run-off from these facilities.

The following practices are recommended:

- District Maintenance Director or Environmental Personnel should be responsible for ensuring that Facility Pollution Prevention Plans (FPPPs) are developed for each maintenance facility.
- The FPPPs should identify the work activities at each facility along with the corresponding BMPs that should be implemented.
- Supervisors should inspect their maintenance facilities monthly to monitor the implementation and adequacy of the BMPs.
- A report that includes the date of the inspection, the name of the inspector, observations, and recommended corrective actions should be prepared by the Supervisor.
- All inspection records should be maintained for a

period of 3 years. Any observed instances of non-compliance should be reported to the Stormwater Coordinator.

- In addition to monthly facility inspections conducted by the facility supervisor, the more in depth review should occur in at least 20 per cent of each District's facilities each year.
- These reviews should monitor each facility's documentation and include a thorough yard inspection.
- Each District Maintenance Stormwater Coordinator should prepare a report including the date of the inspection, name's, of the inspector, observations, and recommended corrective actions.
- All FPPP records should be maintained for a period of 3 years by the Maintenance Supervisor.
- Any observed instances of noncompliance should be reported in accordance with procedures.
- In addition to inspections conducted by the facility supervisors DOTs may employ an audit programme or other supplementary compliance monitoring to support continual improvement.

ENVIRONMENTAL INFORMATION SYSTEMS

Environmental management systems are increasingly used by state DOTs to avoid generation of pollution and manage operations for continual environmental improvement. Examples follow, some of which are described in greater detail in sections 2.5, Measuring Environmental Performance and 2.6, Environmental Staffing, Roles, and Responsibilities. Short of an EMS, a number of DOTs conduct surveys of all maintenance facilities to establish compliance with federal, state and local environmental regulations.

Maine DOT's EMS for Facilities

Maine DOT has developed and implemented Environmental Management Systems for all MDOT facilities.

Combined Environmental and Office of Health and Safety Administration (OSHA) policy and procedure manuals are targeted to the managers who have responsibility for implementation. Quick reference environmental practice guides-written as a companion guide to the policies and procedures-were developed for supervisors and field crews.

MDOT's commitment to conduct annual audits of its facilities to systematically review the effectiveness of these policies and procedures has been an important aspect of implementing new environmental procedures. An Environmental Management Committee is responsible for tracking and timely closure of audit findings and development of a database of Corrective Action Reports. MDOT's audit programme and performance measures are discussed in the respective sections of this report.

Massachusetts Highway's EMS for Facilities Management

Mass Highway's EMS for Facilities Management focuses on hazardous waste and hazardous materials, underground storage tank management, wetland and water quality protection, and solid waste management. System Improvement and Implementation plans are developed for each facility.

Mass Highway has developed an implementation manual describing organizational roles and responsibilities relative to environmental compliance management at Mass Highway facilities. Personnel within the major Organization Offices, Divisions, Districts, and Sections that affect compliance with Mass Highway environmental requirements are identified, along with associated training programmes to educate staff "how to best carry out their environmental related duties."

Environmental Management Programme

Maintenance District 10 developed Process Maps operations associated with each significant aspect of operations with a special focus on the District 10 Maintenance Facility, providing information to plan, conduct, assess, and complete activities according to "Plan-Do-Check-Act" framework and principles.

Process Maps identify responsibilities associated with each action. For example, PennDOT staff developed Quality Assurance Evaluations for Maintenance Stockpiles and Foreman's 15-Minute Stockpile Walkarounds.

PennDOT implemented procedures to enhance environmental performance, including annual calibration of spreaders before the onset of the winter services season, use of two-way radios between operators during storms to communicate information about application rates and roadway temperatures, daily electronic leak detection tests in the morning hours before the day shift at garages with corrective action if necessary to prevent leaks, and completion of a Foreman's Erosion and Sedimentation Checklist as part of planning for earth disturbance activities that require control measures.

PennDOT's ISO-based SEMP plan resulted in:

- Development of information on contractor/supplier procedures and requirements related to significant aspects, which are consistent with department-wide contract terms and conditions, requirements, and procedures.
- Establishment of procedures for emergency response and spill prevention.
- Development of procedures, checklists, and responsibilities in monitoring and measurement activities related to significant aspects.
- Internal development of auditing procedures for SEMP activities performed by trained staff from another district.

New Hampshire DOT's Inventory of Managed Properties

New Hampshire DOT (NHDOT) developed an IMP (Inventory of Managed Properties) to inventory hazardous materials at all of the Department's maintenance and operations facilities. NHDOT operations policy requires that all NH DOT properties be screened and all environmental concerns recorded. To address this need, NHDOT implemented a system of integrated handheld computers and web-based data management to support a contaminated property valuation policy for prospective and currently owned properties.

NHDOT's increased emphasis on inventory and risk management of all properties potentially impacted by a project or currently owned by the state greatly increased the volume of hazmat data being collected and managed. It also placed more emphasis on early detection. A robust site screening protocol was developed to collect preliminary field observations of hazmat sources and receptors. To support the protocol, the field data collection application was developed for use on personal digital assistants (PDAs). The PDA software standardizes site-screening data, improves data completeness and quality, and reduces time delays from fieldwork to data reporting.

Since digital photographs and GPS data are captured using integrated hardware, and are stored directly to the database upon collection, there is no sorting, labeling, and management of this information following field work. The database is dynamically linked to the Bureau of Right-of-Way, ensuring that property information is kept accurate and redundancy of data is eliminated. Functionality built into the graphic user interface on the web calculates "risk scores" for each property and prioritizes all of the sites within a corridor, flagging key hazmat issues.

The developed technology provides the NHDOT with better and faster data from the initial phases of a project; the ability to "triage" sites based on their calculated risk rankings and flags; and the capability to manage contaminated sites from identification through remediation within the web application. IMP also allowed NHDOT to easily communicate with the state's Department of Environmental Services (NHDES). Minor incidents do not need to be reported directly to the department, as long as the occurrence is posted on the database, eliminating several sets of paperwork, which would normally need to be prepared for both NHDOT and NHDES.

This new technology has reduced the time spent on site, and standardized data collection and reporting performed by consultants. Currently, IMP is used solely in the documentation of hazardous waste inventories, remediation and issues for each of the DOT maintenance facilities throughout the state; however, it will soon be used to document stormwater management and cultural resource issues at all DOT owned sites. In conjunction with IMP, NHDOT also developed a Risk Assessment for Site Contamination and Appraisal of Lands (RASCAL). Though developed primarily for project development and right-of-way purposes, it is also used by construction personnel to determine the status of hazardous materials cleanup at construction sites.

FACILITY SITING AND PRIORITIZATION OF ENVIRONMENTAL IMPROVEMENTS

Facility Siting Considerations

Currently, future sites for DOT facilities are usually selected based on cost of land acquisition and operational convenience. Some facility sites have been acquired through "swapping" an existing DOT site for a more desirable parcel. Environmental factors are often not considered and evaluated, unless a procedure specifying such consideration is in place and/or information has been made readily available or a study has been performed.

Information on existing DOT maintenance facilities is needed to allow identification and ranking of sites that are the most environmentally sensitive, to decide which sites to address first. Such information enables DOTs to:

- Prioritize sites that should be closed or relocated based on environmental concerns, as funds become available or on a more pressing basis.
- Identify facilities that require pollution control devices, such as oil/water separators or implementation of other environmental stewardship practices, and those that need to implement stormwater run-off controls.
- Identify environmentally appropriate locations for new facilities, including newly-designed salt storage buildings.
- Develop and implement appropriate decommissioning

policies or procedures. Many DOT maintenance facilities are currently closed and/or relocated without a decommissioning policy or procedures. This can result in abandoned areas of actual or potential contamination and/or the transfer of hazardous and non-hazardous chemicals and wastes to other DOT facilities without advanced planning and, sometimes, without advance notification.

Environmental Data Needed for Evaluation in Facility Siting

Consideration of the following widely available environmental data is recommended in considering facility siting and future changes that may be needed to improve environmental stewardship.

Most of this data is available from state environmental quality or natural resource agencies, or a federal agency if noted:

- Well log data
- Soil borings
- Surface water intakes and wellhead protection areas for public drinking water systems
- High-volume groundwater users
- Spill, Superfund, Leaking Underground Storage Tank and other contaminated sites locations of groundwater aquifers and surface water bodies (EPA)
- Environmentally sensitive areas. Criteria used by INDOT for identifying "sensitive waters" include those waters:
  - Providing habitat for s pecies of concern; having state or federal designations of endangered, threatened, rare, extirpated or on a "watch list" identified by generic descriptor or heritage species code.
  - Used as a public surface water supply intake; maintenance facilities are within 1,000 feet, 3,000 feet or one mile of a public water intake.
  - Used for public recreation; within a mile of such a recreation area and not connected to a POTW.

  - Classified as outstanding state resource waters or high quality waters
- Groundwater aquifers and surface water bodies.
- Locations of urban wet-weather and rural (agricultural) drainage patterns.

For maintenance facilities that are captured under the Municipal Separate Storm Sewer System (MS4) portion of the NPDES programme, the DOT is required to assess the water quality of known receiving waters and stormwater outfall discharges and known sensitive areas, and to identify those places having a reasonable potential for causing stormwater problems.

In case of the latter, DOTs are expected to implement control measures and conduct operations in ways that will reduce contamination of stormwater discharges.

As a result, it is important to:

- Identify facilities that are not currently connected to a Publicly Owned Treatment Works (POTW)
- Attempt to connect to a POTW when new sites are developed.

Criteria for Prioritizing Attention to Maintenance Facilities

Utilizing data such as that discussed above, criteria can be developed to identify maintenance facilities that should receive priority attention.

INDOT utilized the following criteria that are applicable to other states, to identify those that provide the greatest potential risk to the environment from stormwater discharge, locations both within and outside MS4 areas:

- Maintenance facility locations within designated MS4 areas.
- Maintenance facility locations within 3,000 feet of a community public well.
- Maintenance facility locations within (1,000 feet) (3,000 feet) (5,280 feet) of a public surface water intake.
- Maintenance facility locations within one mile of high quality and exceptional use waters.

- Maintenance facility locations within one mile of federal, state, county, municipal or township recreation facility having a lake, pond, river, or stream.
- Maintenance facility locations within 3,000 feet of groundwater that is highly vulnerable and very highly vulnerable to contamination by nitrates (as surrogate for chloride).
- Maintenance facilities within 3,000 feet of a natural area containing Rare, Threatened, or Endangered species.
- Maintenance facilities within one mile of the "best remaining examples of natural wetland communities," as determined by IDNR.

FACILITY HOUSEKEEPING PRACTICES

Daily activities occurring at maintenance facilities can involve the use of materials and products that are potentially harmful to the environment. Many DOT "yards" or "depots" are the location of aggregate piles, metal scrap piles, miscellaneous right-of-way trash, and other debris that can potentially contaminate stormwater. Stormwater run-off has the potential to come in contact with and transport sediment and other pollutants from the facility grounds to storm drains or adjacent water bodies. Non-stormwater, from sources such as landscape watering, vehicle cleaning, water line/hydrant flushing, and air conditioning condensation, can also transport pollutants as it flows across facility grounds. Good housekeeping practices are intended to eliminate the potential for discharge of pollutants to drainage paths, stormwater drainage systems, or watercourses by promoting efficient and safe storage, use, and cleanup of potentially harmful materials. The best strategy for minimizing pollutants in discharges from the facility is to control pollutants at the source.

GENERAL STORMWATER PROTECTION PRACTICES AT DOT MAINTENANCE FACILITIES

Stormwater and non-stormwater can be prevented from

coming into contact with potential pollutants by use of the following practices, outlined by Caltrans in their bulletins for maintenance staff:

- Cover stockpiles and other materials stored outdoors.
- Use berms or other containment methods to prevent run-off.
- Sweep paved areas to remove sediment and other materials that have been tracked or dispersed across the facility.
- Ensure that paved surfaces are in good condition.
- Prevent non-stormwater, such as condensate water from ice machines and sprinkler overspray, from flowing across facility grounds.

BMPs should be installed at storm drain inlets, catch basins and facility discharge points as final defence measures in the event preventive measures are not fully effective.

Since spills and leaks may occur at any time, preparation should be in place, including the following practices:

- Locate raw material stockpiles away from drain inlets and catch basins.
- Do not repair, maintain, or clean vehicles and equipment near inlets.
- Move receptacles, hazardous waste areas, raw materials storage areas, vehicle wash areas, and stockpiles away from drain inlets and areas that are prone to flooding or ponding.
- Do not park vehicles and equipment over or immediately adjacent to inlets.
- If a spill occurs, clean up the area immediately and dispose of cleanup materials properly.
- Stencil drain inlet locations with paint or signs.
- Maintain sufficient emergency materials; such as drain covers, absorbent booms, rags, or sandbags convenient to inlets.
- To prevent flooding, place BMPs so that water will drain while retaining the pollutant on site.
- Inspect culverts, ditches, gutters, underdrains,

horizontal drains, downdrains, and outlets annually, and as needed during the rainy season, to determine if cleaning or repairs are needed. This prevents the drainage structure from becoming a pollutant source itself.

- Collect and manage all water and material generated during drainage facility cleaning operations per solid and liquid waste management practices.

Caltrans recommends the following maintenance yard housekeeping practices in their statewide stormwater quality practice guidelines:

- Provide facilities for containment of any accidental losses of concentrated solutions, acids, alkalies, salts, oils, or other polluting materials.
- Employ standard operating procedures for spill prevention and clean up during fueling operations, as well as BMPs for vehicular maintenance areas.
- Prohibit equipment or vehicle wash waters and concrete or asphalt hydrodemolition wastewaters from flowing into stormwater run-off, except under an appropriate NPDES wastewater permit.
- Promote recycling and manage solid waste according to the appropriate procedures or stewardship practices.
- Minimize pesticide, herbicide and fertilizer use. Pesticides should be used, applied, handled, stored, mixed, loaded, transported, and disposed according to manufacturer's procedure and any state requirements.
- Use the "first in first out" policy for material storage and control. Avoid ordering more materials than can be stored properly or used in a reasonable timeframe. Properly reuse, recycle, or dispose of empty containers, excess materials, equipment, and parts that are not likely to be used.
- Clean up spills promptly.
- If it is necessary to use a hose for cleaning, wash water

should not be discharged to the stormwater drainage systems or watercourses.

BUILDING AND GROUNDS MAINTENANCE

Permanent maintenance facilities require building and grounds maintenance, which includes care of landscaped areas around each facility, cleaning of parking areas and pavements other than areas of industrial activity, and maintenance of the stormwater drainage system. Tasks to perform these activities include equipment operation, litter/trash pickup and maintenance of restrooms/RV dump stations and landscaping, which can in turn result in spills, leaks, trash, sewage, erosion and chemical vegetation control. Potential pollutants include litter, trash, sewage, pesticides, fuel, hydraulic fluid and oil.

Recommended environmental stewardship practices include:

- Maintain equipment and buildings to avoid peeling paint, rust, and degradation. Request funding for major repairs.
- Maintain clean, orderly material and equipment storage areas. Provide covers for materials as needed.
- Sweep or vacuum maintenance facility floors and pavement.
- If mopping is used to clean floors or pavement, contain the mop water and dispose of it to the sanitary sewer system according to the following guidelines:
- Do not dispose of mop water into the parking lot, street, gutter or drain inlet; and
- If an oil/water separator is available, pour the mop water into the separator so that the wastewater is treated before being discharged to the sanitary sewer system.
- Minimize the possibility of stormwater pollution from outdoor waste receptacles by doing at least one of the following:
- Use only watertight waste receptacle's, and keep the lid's, closed;
- Grade and pave the waste receptacle area to prevent run-on of stormwater;

- Install a roof over the waste receptacle area;
- Install a low containment berm around the waste receptacle area; or
- Use and maintain drip pans under waste receptacles.
- Utilize the following environmental stewardship practices to protect water quality: scheduling and planning, illegal spill discharge control, safer alternative products, vehicle and equipment fueling, vehicle and equipment maintenance, sweeping and vacuuming, silt fence, sandbag and gravel bag barrier, straw bale barrier, fiber rolls, wood mulch, compaction, spill prevention and control, solid waste management, liquid waste management, sanitary/ septic waste management, hazardous waste management, concrete waste management, material delivery and storage, material use, litter and debris, potable water/irrigation, water conservation practices, maintenance facility housekeeping practices, and compaction.

VEHICLE AND EQUIPMENT MAINTENANCE

The following stewardship practices apply to equipment maintenance:

- Maintenance should be performed in covered or indoor maintenance areas where potential pollutants cannot be introduced into stormwater drainage systems.
- Inspect equipment for damaged hoses and leaky gaskets and repair or replace as necessary.
- Drip pans or absorbent materials should be used during vehicle and equipment maintenance work that involves fluids.
- Non-stormwater discharges into stormwater drainage systems or watercourses are prohibited.
- Utilize Spill Prevention and Control BMPs for pollution prevention and response measures. Any contaminated soil resulting from vehicle or equipment repair should be addressed.

- Use dry methods for cleaning associated with maintenance in outdoor areas.
- Inspect areas following field maintenance areas to ensure there is no residual contamination that might impact stormwater quality. Clean areas as needed using dry methods.
- Maintain waste fluid containers in leak-proof condition.

Pre-Operation Inspection

- Vehicles and equipment should be inspected for leaks on each day of use. When performing pre-operation inspection, pay particular attention to:
- Ensure that the vehicle/equipment is clean and in good operating condition.
- All equipment has current inspection stickers, as applicable.
- Assignees and operators of motorized vehicles and equipment ensure that preventive maintenance occurs and that all malfunctions, operating problems, etc., are reported for corrective action. Request the repair of vehicles/equipment with leaks. Place a drip pan under any leaking vehicle or equipment. Preventive maintenance should occur in accordance with departmental guidance.
- Clean up spilled or leaked fluids immediately.
- Verify that hoses and clamps are secure and check for evidence of leaking.
- Problematic vehicles or equipment should be removed from the maintenance activity site.
- Daily pre-trip inspection should be logged and kept for 3 months.

Vehicle Fluid Removal

When removing automotive fluids such as used motor oils, coolant, or other oils from vehicles or equipment, the following environmental stewardship practices should be used:

- Transfer removed fluid to a designated used fluid storage tank as soon as possible.
- If possible, remove fluids directly into the holding tank. For example, newer types of used oil tanks can be connected to the vehicle to pump oil directly into the tank.
- If necessary, drain fluids into a drip pan and then transfer the fluids to the designated container. A larger drip pan may be required to catch any unanticipated splashing.
- Properly remove, clean, and store drip pans promptly after use.

Engine and Parts Cleaning

When cleaning engines and parts during vehicle and equipment repair operations, the following environmental stewardship practices should be used:

- Designate specific areas for parts cleaning.
- All parts washing should be performed in designated areas with captured wastewater.
- Use self-contained sinks or tanks when working with solvents. Periodically check for leaks and make necessary repairs as soon as possible. When not in use, make sure covers are secure.
- After rinsing parts, allow them to drain and dry over a solvent sink or tank. This will prevent dripping onto the floor.
- All vehicles and equipment should be washed at an approved area.

Cleaning Up Spills of Vehicle and Equipment Fluids

Accidental releases of vehicle fluids at maintenance sites can potentially discharge into stormwater drainage systems and pollute receiving waters. Typical vehicle fluids include oil and hydraulic fluids leaking from vehicles and equipment, accidental spills from fueling operations, and leaks and spills around storage tanks and containers.

Caltrans developed and distributed the following environmental stewardship practices for cleaning up spills of vehicle and equipment fluids:

Proper response to a vehicle fluid leak requires preparation:

- Maintain up-to-date spill prevention, control, and response plans.
- Train staff to identify and respond to spills safely and appropriately.
- Maintain appropriate and adequate supplies of cleanup materials at fueling areas, vehicle maintenance areas, cleaning areas, and vehicle and equipment parking areas.
- Regularly inspect vehicle parking, maintenance, cleaning, and fueling areas for leaks and spills.
- Repair or replace vehicles and equipment that consistently leak.
- Repair or replace, as needed, material and waste storage perimeter controls, containment structures, covers, and liners in order to contain spills and leaks.

Evaluate the spilled material to determine the appropriate methods for cleaning up the spill. Vehicle fluids such as oil, fuels, and hydraulic fluids are considered hazardous wastes and require appropriate safety precautions. For spilled material, immediately contain the material to keep it from spreading and clean it up.

- Place absorbent materials or pads around leaks to soak up spills.
- For vehicles/equipment that are leaking, place a drip pan underneath to contain any additional leakage.
- Place a leaking container in appropriate spill containment or transfer the contents to another container.
- For leaks or spills that occur during storm events, to the extent that work can be accomplished safely, cover and protect the spilled material from stormwater run-on.

Once the spilled material has been contained, ensure that all of the material and absorbent has been cleaned up.

- Whenever possible, use "dry shop" methods to clean up spills.
- Avoid hosing down the spill area.
- Use an absorbent-type cloth on fuel pumps or damp mop on pavement in fueling areas.
- If rainwater has accumulated in a contained area where a spill or leak has occurred, the contaminated water might be considered hazardous waste.
- Take additional precautions in situations where dry cleanup methods cannot be implemented to ensure that the water used for cleaning and decontamination is prevented from entering storm drainage systems or receiving waters.
- Dispose of the contaminated wastes (spilled material) used cleanup materials, contaminated rainwater, according to environmental stewardship practices. Contact the DOT's Stormwater Coordinator or HazMat Coordinator for additional assistance.

SEDIMENT CONTROL AT MAINTENANCE FACILITIES

Sediment on facility grounds comes from two primary sources:

1. Eroded soil from unpaved areas and slopes is transported onto the facility grounds by gravity, wind or water.
2. Mud and dirt are brought onto the facility on vehicle and equipment tires and undercarriages.

Caltrans recommended the following environmental stewardship practices in a bulletin to staff on pollution prevention at maintenance facilities.

Evaluation of Exposed Soil Areas at Maintenance Facilities

Regularly inspect unpaved areas of the facility for signs of erosion and identify factors in selecting appropriate stabilization measures.

- Is the area deficient of vegetation or other conditions or practices to hold the soil in place?

- Is the area subject to run-on, either sheet flow or concentrated flow?
- Does the area have significant slopes that will increase the probability of erosion?
- Is the area being used for equipment storage?
- Do vehicles or equipment regularly utilize these areas?
- Is the area intermittently used for storage of materials or waste?
- Does the area show signs of erosion?

Implementation of Appropriate Erosion and Sediment Control at the Facility

- Maintain existing vegetation and enhance where possible.
- Prevent run on from adjoining areas that can cause erosion using ditches, berms, dikes, or swales, sandbag or gravel bag barriers, or fiber rolls.
- Protect slopes, flat areas, exposed soil areas, or transportation corridors with gravel or pavement, if possible, otherwise use applicable BMPs that best fit the facilities needs, such as wood, straw or hydraulic mulch; seeding; or compaction. Well-maintained mulch provides cost-effective erosion control benefits.
- Do not over-irrigate landscape vegetation. Ensure irrigation systems are in proper working order and not over watering or overspraying areas.
- Inspect unpaved/disturbed soil areas regularly to assure that erosion and offsite sediment discharge is not occurring.
- Minimize use of chemicals to eradicate vegetation from exposed areas.
- Prevent storage of hazardous materials on exposed soil areas.

Inspection and Cleaning

For the most effective programme to reduce sediment and raw materials in stormwater, a routine inspection and cleaning programme is needed with the following elements:

- Regularly sweep or vacuum the facility grounds to remove accumulated pollutants.

- Regularly inspect drop inlets, facility discharge points, and facility perimeters, for accumulated pollutants. Remove pollutants and implement BMPs as indicated.
- As indicated by the inspection, implement linear sediment barrier controls, silt fence or gravel bag barrier, etc.
- Maintain sediment controls by removing accumulated sediment and repairing damaged areas as required by the BMPs.

Regularly inspect facility vehicles and equipment for dirt and mud. Ensure that vehicles are cleaned at designated washing facilities.

MATERIALS MANAGEMENT AT MAINTENANCE FACILITIES

Maintenance facilities store a variety of products that may be harmful to the environment if they come into contact with surface waters.

Materials that may be stored include pesticides, petroleum products, paints, cement and solvents, as well as bulky items such as:

- Brush, wood, and untreated lumber
- Treated lumber, poles, ties, including pressure treated or creosote treated
- Scrap metal and obsolete machinery
- Construction materials stored for reuse (culverts, steel beams, guard rail, cable, etc.)
- Old tires, cardboard, signs, sign posts, plastic, bulky trash
- Dirt, road sweepings, ditching material, inert fill, sand and gravel
- Rock or stone
- Old bricks, concrete, or asphalt

Potential pollutant sources are contaminated run-off as well as spills and leaks that may release pesticides, paint, solvents, asphaltic products, cement, epoxy resins, fuel, hydraulic fluid, and oil. Good storage and handling practices can greatly minimize waste quantities and costs for disposal as well as reduce potential for employee exposure or environmental contamination. Environmental stewardship practices can significantly reduce handling, disposal costs, and future liability from DOT activities. Division Managers/Engineers are ultimately responsible for maintaining all Maintenance and Operations properties in good order, for minimizing the amount of material stockpiled at maintenance lots, and for ensuring that materials are stored, reused/recycled, or disposed in accordance with this procedure. Crew supervisors are responsible for ensuring that maintenance crews know and understand the procedure for storing and disposing of bulk materials.

Stockpile and Materials Management Practices and Procedures

The following items should be considered in the development of a stocking area:

- Blacktop Pad of sufficient size to accommodate materials storage, loading and mixing materials if required. Also add curbs on a paved swale section to channel run-off into collection basins. All bituminous surfaces should be sealed to make the surface impervious.
- Collection Basin must be properly placed for easy cleaning and effective functioning.
- Permanent Covered Chemical Storage Building.
- Lighting sufficient for loading area plow mounting area, etc.
- Truck Heaters—spaced in a row, not bunched on a single pole.
- Grading and Access Roadway Work.
- Signing stockpile identification, Maintenance.
- Fencing and Security.
- Planting for Environmental Screening.
- Identification of Stocking Area Boundaries.
- Types of Bulk Storage Buildings ranging from barn type buildings storing a few hundred tons of bulk chemicals to large diameter dome buildings with storage capacities over 5,000 tons.

- Location. Columbia County in Wisconsin placed their new salt storage domes in a location that allowed the building of 1,800 ft of railroad-side track. This allowed bringing in the annual salt requirement by train, thus reducing the shipping cost greatly versus trucking. The cost of salt was decreased from $33 a ton to $22.34 a ton.

Environmental stewardship practices and procedures in materialsmanagement include the following:

- Material should be recycled when possible. When waste materials are not recycled such materials should be disposed in accordance with regulations and/or department policy.
  - Untreated wood waste and brush should be chipped and reused as mulch or fuel.
  - Scrap metal should be recycled
  - Treated wood should be reused when possible or sent to a licensed disposal facility permitted to accept this material. If new creosote treated timber is being temporarily stored for more than a week it should be covered and lined underneath with an impervious material. The liner should be bermed to catch any leachate. The leachate should be cleaned with absorbent pads that can be disposed of as oily rags. Old, used timber may be stacked without cover or liner.
  - Develop site plans for areas adjacent to or near riparian areas to identify erosion and sediment control needs, and to ensure stability of the material.
  - Sites should be identified as part of the local disposal plan.
  - All stockpiles should be located away from concentrated flows of stormwater, drainage courses, and inlets.
  - All stockpiles should be protected from stormwater run-on, using berms, dikes or other temporary diversion BMPs.

- Maintenance facilities should appear generally clean and well organized.
- Designate specific areas for temporary stockpiling of various types of bulk materials and wastes, such as scrap metal, brush/wood, old signs/lumber, or inert fill.
  - Signs, fencing, site plans or other markings should be used to identify stockpile areas.
  - The total bulk and waste material storage area not exceed limits specified by the DOT or state environmental agency.
  - Inspect and organize the storage areas, particularly before rainier seasons. Remove litter, debris, sediment, and any spilled materials to prevent potential pollutants from being introduced into stormwater run-off.
  - Store materials away from areas that have potential for run-off into the stormwater drainage system or other watercourses.
  - Where feasible, cover materials that may have potential to impact stormwater quality during the rainy season. For materials that are frequently used, keep covers or tarps available for use during rain.
  - Frequently sweep around storage areas to remove materials blown, tracked, or washed onto surfaces that may wash off with rain.
  - Clean any spills or drips collected in secondary containment and spill containment facilities for above ground tanks and other storage/waste containers to prevent contamination of collected stormwater. Drain plugs and valves should be secure.
  - Clean vehicle wash rack sumps, clarifiers, and oil/water separators exposed to rain, as needed, to ensure free drainage and to prevent possible overflow.

- If debris, sediment or other materials still have the potential for impacting stormwater run-off even though source controls are in place, consider installing temporary sediment controls (sand bags) straw bales, filtration socks, etc., at inlets, stockpile areas or other sources. Make sure inlet protection will not contribute to flooding. Remember, inlet protection is intended as secondary protection only and may not be needed if source control BMPs are in place.
- All deployed BMPs should be inspected regularly during the rainy season, particularly before and after rain events. Inspecting BMPs during rain events can be beneficial in determining their effectiveness and identifying any needed modifications. Re-inspect inlets and drainage facilities after rain events. Clean and repair as necessary to ensure that drainage facilities are functioning properly.

a. Sediment Controls - Inspect sediment controls such as sand bags, straw bales, silt fencing, and sediment traps and basins. Remove captured sediment from the sediment controls before the rainy season. Replace or repair degraded sand bags, straw bales, or silt fencing as necessary.

b. Drainage Facilities - When inspecting drainage facilities take note of their condition along with the condition of any associated BMPs. If excess sediment, debris, or other potential pollutants are observed in or near the drainage facility, look upstream at the sources and consider modifying or implementing additional BMPs. If needed, implement temporary drain inlet protection.

- Stockpiles should incorporate erosion and sedimentation controls and prevent erosion or sediment discharge into rivers, streams, ponds or wetlands. Caltrans has recommended the following stewardship practices for control of sediment from raw material storage areas.

- Water quality, erosion and sediment control BMPs should be properly implemented and regularly maintained. Interim sediment controls include using temporary sediment controls such as sand bags, straw bales, or silt fences to contain raw materials. Temporary sediment controls, such as sand bags and straw bales can degrade and may contribute to stormwater pollution. Temporary and permanent sediment controls should be inspected regularly and replaced or repaired as needed. Sediment contained by temporary or permanent controls should be removed periodically.
- Wind erosion control practices should be implemented as appropriate on all stockpiled material.
- In general, stockpiles should be covered or protected with a temporary perimeter sediment barrier at all times. Perimeter controls and covers should be repaired and/or replaced as needed to keep them functioning properly.
- Berms should be installed around storage areas to minimize tracking of materials out of storage areas and to contain sediment within the storage area. Permanent rolled berms or ramp berms should be made of hot asphalt or Portland Concrete Cement (PCC). Cold mix asphalt is not recommended for use as raw material containment berms. Over time, cold mix has the potential to break up and not function as well as hot mix asphalt or PCC. Cover raw materials (especially cold mix) during the rainy season and have covers readily available outside the rainy season when rain is predicted.
- Sweep surfaces where material is tracked, blown, spilled or washedfrom the storage area.
- Reduce the size of stockpiles or the amount of stockpiled materials during the rainy season.

- Material that has been contaminated with oil, gasoline or other chemicals should not be used as fill. Any material suspected of contamination should be reported promptly.
- Environmental staff should be called for assistance with materials placement or permitting issues as needed.
- Approved areas for filling should be marked by stakes or other markings, and appropriate erosion and sedimentation controls should be used.
  - Filled areas should be graded and stabilized by seeding and/or other appropriate methods when filling is complete.
  - Interim or seasonal stabilization should be used if filling occurs over an extended period.
- Stockpiles of scrap metal, wood, brush, asphalt, or waste materials having no future use should be completely removed at least annually. In addition to minimizing environmental impacts, this will help avoid having the site be considered a solid waste disposal facility.
- Obsolete equipment being stored for salvage or parts should be stored in a designated area and protected from weather, as appropriate. Fuels, oil, and fluids should be removed or properly contained to prevent spills or leaks.
- No material should be disposed of or buried on maintenance lots, except inert fill or other authorized material (such as deer carcass composting in the case of NYSDOT). Avoid burying or disposing of:
  - Old drums or containers
  - Chunks of hardened calcium chloride or sodium chloride
  - Paint, paint containers, fuels, oils, or other hazardous materials
  - Rubbish or garbage
  - Pesticides
  - Old culverts

Salt and Sand Stockpile Management

Soil and water contamination may occur around the salt sheds or sand piles if poor housekeeping practices are in place.

Recommended environmental stewardship practices include the following practices compiled from Iowa DOT, PennDOT, NYSDOT, Missouri DOT, the Transportation Association of Canada, and the Alberta, Canada Transportation Authority:

- All storage facilities should be inspected and repaired regularly for roof leaks, floor cracks and wall leaks.
- All stored material is under roof, on impervious pad, in areas properly sized for truck and loader operations, stocked below fill line. Piles of salt are not left exposed to the elements.
  - Salt and mixtures of salt and sand are kept on an impermeable surface like asphalt or concrete and in salt storage buildings whenever possible.
  - Under some circumstances, temporary "surge" piles may be utilized if placed on an impermeable surface and covered with adequate (weighted) tarping.
  - Doors to the salt sheds and sand domes are kept closed unless salt is being delivered or removed. Keeping the door closed ensures that the salt remains in the shed, away from snow and rain. Material must be tarped within ten feet of doorway. Maintenance staff at Iowa DOT designed and installed an innovative but basic canvas salt shed door that lifts easily, allows for full access, and provided substantial cost savings.
  - Where fabric buildings can withstand winter snows, such structures have offered one of the most cost effective methods to keep salt under cover and provide winter storage of the mixes and other de-icing materials. Missouri DOT has found such buildings to be durable, low cost, spacious and able to be installed on a permanent or

temporary foundation. Such buildings have provided storage space 2,000 to 3,000 tons, with room to work inside. Salt run-off has been eliminated at the storage sites.

- With Bay Storage Bins and Crib Storage, the front of a barn storage bin is open and when the building is full, the salt is partially exposed. Therefore, the following environmental protection items must be followed to guard against leaching and run-off. These environmental stewardship practices are also necessary for crib storage, to guard against leaching and run-off, as crib storage is not roofed:
- The bituminous pad on which the building is placed must extend for a distance of 20 feet past the front of the building.
- The building is not to be overloaded so that salt spills out past the front of the building.
- When fully loaded, the front of the salt pile is to be covered by tarpaulins.
- A sedimentation basin must be constructed to collect run-off.
- The immediate area around the building is to be kept clean of salt spillage that will normally occur when loading the building with trucks. This is especially important for the pad surface in front of the building.

- The area must be properly signed.
- Liquid De-Icer (Magnesium Chloride, Calcium Chloride and/or IceBan/MAGic) etc., which are not included under the Chemical Bulk Storage regulations, is stored in aboveground storage tanks (typically 3,000 - 5,000 gallon).
- Liquid De-Icer storage tanks are located on level compacted sand bases and protected from traffic by barriers. A basic rule of thumb to determine storage needs is 1.5 times total lane miles to treat x

recommended gallons per lane mile = amount of storage (ex: 1.5 x 200 x 50 = 1,500). Iowa DOT normally purchases 2500 gallon storage tanks because their cost per gallon is considerably less than other storage tanks. Other size tanks are available for limited space needs.

- Area drainage is such that any spills can be contained on site.
- Placards or stenciled lettering are used to identify liquid de-icer tank contents.
- Drains must be closed.
- A minimum of two Stockpile Quality Assurance Evaluations should be completed per year by District Offices, one in summer and one in winter. QA is performed by Central Office. Each item receiving a score of 3 or less requires a Correction Action Report (CAR) to be completed and entered into a District tracking system to assure improvement is made.
- A "Stockpile Snapshot" is a cursory stockpile review that can be completed by anyone from the District Office. Any deficiency noted should be addressed within two weeks. A Foreman's Stockpile Checklist is completed by the assigned stockpile Foreman four times per year and reviewed. A copy is sent to the District Office Maintenance Unit and the Facility Administrator. PennDOT awards Silver and Gold Awards to County Maintenance Organizations for Model Stockpiles meeting certain criteria. If all five are met, a Gold Award is given. If four are met a Silver Award is bestowed. An award for "Most improved" is given as well.
- Spills are cleaned up immediately, using necessary equipment.
- If salty water from the stockpile is caught in a holding pond, the pond must be able to contain the amount of water from the next normal storm. It should be pumped down to ensure that this level can be

maintained. The pond water levels should be monitored and excess salt water should be disposed of in an approved location.

- Any contractor activities at government-owned facilities should be monitored to ensure that they are following the operating plan for that facility.
- Operating plans should be developed by maintenance staff and the contractor, if appropriate, in conjunction with DOT environmental staff.

PENNDOT SALT STOCKPILE MANAGEMENT

PennDOT provides winter materials storage, run-off control, training and quality assurance, with a high emphasis on stockpile management.

PennDOT's Model Facilities Task Force (MFTF), comprised of various representatives from Facilities Management Division of the Bureau of Office Services, Bureau of Maintenance and Operations, Bureau of Environmental Quality, Engineering Districts and County Maintenance organizations, indicated a need for PennDOT to reemphasize stockpile management after finding safety deficiencies, improper handling and storage of materials, environmental remedial costs, building damage, and failure to update and implement Preparedness, Prevention and Contingency (PPC) plans.

Now, a District-approved, stockpile-specific PPC plan is displayed unobstructed in the staging building and is revised annually.

PennDOT uses a 50 element QA review, each tied into a department policy or regulation or a PennDEQ regulation. PennDOT has developed a Stockpile Academy Training Programme, which maintenance staff are required to attend on a 4-year rotating basis.

PennDOT inventories winter materials and transfers all environmentally sensitive materials to permanent material storage buildings should begin, starting with any stockpiles located within 500 feet of any wells or streams. PennDOT is moving towards the goal of every county having all salt under roofed storage from May to October. Bins and storage buildings, collection basins, and storage pads are cleaned and repaired in the spring. Prompt spring clean up of anti-skid materials prevents clogging of drains and impairment of surface waters and habitats.

PennDOT requires ongoing evaluation stockpile housekeeping measures, including a list of quality assurance responsibilities, which is included in the The procedure for this checklist includes:

- Completing the checklist for each stockpile by November 30, January 31, March 31 and June 30 of each year.
- The completed checklists are forwarded to the responsible Assistant Maintenance Manager for their review and signature.
- The Assistant Maintenance Manager forwards the signed checklist to the County Maintenance Manager.
- Within ten days of the completed checklist date the County Maintenance Manager forwards all Stockpile Checklists for his/her county to the Assistant District Engineer/Administrator-Maintenance (ADEM/ADAM) and the District Facilities Administrator (FA).
- The FA will determine appropriate corrective action in cooperation with the ADE/A-M, the County Maintenance Manager, Equipment Manager and Assistant Maintenance Managers.

PennDOT makes use of the following quality assurance evaluation indicators for solid winter materials stockpiles:

- Any salt, mixed or treated material not under roof or tarped and anchored with sand bags; or not on an impervious pad.
- Any bagged deicing chemicals not stored on pallets and either under roof or 100 per cent covered by tarps and anchored with sand bags.
- All salt, mixed or treated material stored under roof or on an impervious pad, tarp covered and anchored with sand bags. Note: Tarp and sand bags are not

required during general snow and ice control operations. Bagged deicing chemicals stored on pallets and 100 per cent covered by tarps and anchored with sand bags.

- All salt, mixed or treated material stored under roof, on an impervious pad, below building fill line, and tarp covered and anchored with sand bags. Note: If face of material is more than ten feet from the building doorway, no tarp is required.
- Bagged deicing chemicals stored on pallets and 100 per cent covered by tarps and anchored with sand bags.
- All salt, mixed or treated material is stored under roof, on an impervious pad, and below building fill line, and tarp covered and anchored with sand bags. Note: If face of material is more than ten feet from the building doorway no tarp is required. Bagged deicing chemicals stored under roof, on pallets.

As a result of their system, PennDOT has been able to work with PennDEP to have one permit per district with an EMS in place, rather than one permit per stockpile. PennDOT is not required to sample because an EMS and BMPs are in place and salt is stored under cover. Finding covered loading was not considered necessary because loading areas are paved, curbed, and contained.

PennDOT developed the 15-minute Stockpile Walkaround to be performed by the Maintenance Foreman, along with a shorter Stockpile Snapshot as shown in the Appendix. Maintenance stockpile activities have been charted in a Stockpile Activity Protocol Matrix, also listed in the Appendix.

Material Delivery and Storage

Material delivery and storage procedures and practices are designed for the proper handling and storage of materials at the maintenance facility.

Such materials may include aggregate, pesticides, fertilizers, detergents, plaster, petroleum products, asphalt and concrete components, hazardous chemicals, concrete compounds, or other materials that may be detrimental if released to stormwater drainage systems or watercourses.

The following procedures and practices minimize or eliminate the discharge of these materials to stormwater drainage systems or waters of the state.

- During the initial stocking and following deliveries, special care should be taken to load and pile all solid materials in the approved manner and keep storage locations neat and orderly.
- Store drums in protected (dry) and temperature-compatible manner. Do not store materials that can freeze in unheated areas.
- Containment facilities should provide for a spill containment volume equal to 110 per cent of the largest container in the facility.
- Liquids, petroleum products, and substances listed in 40 CFR Parts 110, 117, or 302 should be stored in approved containers and drums and should not be overfilled.
- Containers and drums should be placed in temporary containment facilities for storage. A temporary containment facility should provide for a spill containment volume able to contain precipitation from a 24-hour, 25-year storm event.
- Containment facilities should be impervious to the materials stored there and maintained free of rainwater and spills.
- Rainwater in containment facilities should be inspected prior to discharge. In the event of soil spills or leaks, accumulated rainwater and spills should be collected and placed into drums. These liquids should be handled as a hazardous waste unless testing determines to be non-hazardous. Nonhazardous liquids should be sent to an approved disposal site.
- Repair and/or replace perimeter controls, containment structures and covers as needed to keep them functioning properly.

Proper Handling and Use

- Personnel at maintenance facilities should be trained to ensure that materials are properly handled and stored.
- Use recycled and less hazardous products when practical, reducing or eliminating use of hazardous materials on-site when practical. Substitute a less hazardous or less waste-producing product or process for those that would otherwise have generated a more hazardous or higher quantity of wastes. As well as potentially resulting in a non-hazardous waste for an indicated activity, such substitution may reduce or eliminate potential employee exposure concerns and additional regulatory burden.
- Use materials only where and when needed to complete the necessary maintenance or construction activity.
- Recycle residual paints, solvents, non-treated lumber, and other materials.
- Do not over-apply fertilizers and pesticides. Prepare only the amount needed. Follow strictly the recommended usage instructions. Apply surface dressings in smaller applications, as opposed to one large application, to allow time for them to work in and to avoid excess materials being carried off-site by run-off.
- Do not apply fertilizers and pesticides to plants that are used by people for food, medicines, basketmaking, and other purposes. Many Native American communities and some other groups use roadside plants for such purposes, and may be harmed by the application of fertilizers and pesticides. Consult with such groups to establish where chemicals should not be applied.
- Identify container contents and maintain data on its contents. Do not remove the original product label from a container, as it contains important safety and disposal information.

- Keep products in their original containers whenever possible.
- Use all of the product before disposing of the container.
- Label any unmarked containers with permanent markers; include the date when filling first occurred.
- Keep a record of what is stored in each container.
- Retain the material safety data sheets (MSDS) for each product.
- Record any information that relates to a waste, such as "also contains some water" or what activity the waste resulted from, such as "Safe-Strip cleaning solvent from epoxy pavement marking activities."
- Whenever possible, return unused products to the supplier.
- Never mix dissimilar materials and wastes in the same containers.
- Drain valves should remain closed except to release clean rainwater.
- Separation should be provided between stored containers to allow for spill cleanup and emergency response cleanup.
- Incompatible materials, such as chlorine and ammonia, should not be stored in the same temporary containment facility.
- To provide protection from rain, bagged and boxed materials stored outdoors should be stored on pallets throughout the rainy season.
- To provide protection from rain, bagged and boxed materials should be covered prior to rain events.
- Storage areas should be kept clean, well organized and equipped with cleanup supplies for the materials being stored. Perimeter controls, containment structures, covers and liners should be repaired or replaced as needed.
- Check to ensure that designated storage areas are kept clean and well organized.

- Dispose of wastes or empty containers regularly.
- Dispose of waste before knowledge of their contents is lost and before deterioration occurs.
- Keep an inventory of the waste on hand and contact environmental staff to set up disposal contracts for both hazardous and non-hazardous wastes.
- Dispose of empty containers promptly before water or other contamination or deterioration occurs.
- Avoid pollution from oil, paint, solvents, diesel, lead-acid batteries, fuel, hydraulic fluid and oil through use of water quality BMPs such as scheduling and planning, vehicle and equipment fueling, vehicle and equipment maintenance, hazardous waste management, material delivery and storage and spill prevention and control.
- Latex paint and paint cans, used brushes, rags, absorbent materials and drop cloths, when thoroughly dry and no longer hazardous, may be disposed of with other construction debris.
- Mix paint indoors or in a containment area. Never clean paint brushes or rinse paint containers into a street, gutter, storm drain or watercourse. Dispose of any paint, thinners, residue or sludges that cannot be recycled as hazardous waste.
- For water-based paints, paint out brushes to the extent practical and rinse to a drain leading to a sanitary sewer, where permitted or into a concrete washout pit or temporary sediment trap. For oil-based paints, paint out brushes to the extent practical and filter and reuse thinners and solvents.
- Keep an ample supply of spill cleanup material near material use areas. Train employees in spill cleanup procedures.
- Spot-check employees and subcontractors throughout the duration of the job to ensure appropriate practices are being implemented.

WASTE MANAGEMENT

Waste management is governed by both federal and state requirements which include Federal hazardous waste management regulations developed under the Federal Resource Conservation and Recovery Act 'RCRA,, state-level waste management regulations, and regulations developed under other statutes that apply to wastewater discharges and air emissions.

DOTs practice stewardship by complying federal and state waste management regulations and taking measures to reduce or eliminate waste streams, in order to reduce the present and future threat to human health and the environment.

PennDOT practices the following hierarchy for waste management decisionmaking:

- First, consider Reduction: Activities that reduce or eliminate the generation of hazardous wastes.
  Waste reduction may include the following benefits:
  - Lower Operating Costs From The Substitution Of Less Expensive Raw Materials.
  - Lower Energy Costs Through The Use Of Newer, More Efficient Equipment.
  - Reduced Transportation And Disposal Costs.
  - Improved Product Quality.
  - Reduced Long-Term Liability Associated With Handling And Disposal Of Hazardous Wastes.
  - Enhanced Employee Safety From Reduced Exposure To Hazardous Materials.
  - Cost Savings From The Reuse Of Materials.
  - Revenues From The Sale Of Surplus Materials.
  - Fewer Regulatory Compliance Requirements.
  - Improved Public Image-The Less Wastes Produced, The Less The Department Is Viewed As A Contributor to environmental problems.
- Second, consider Recycling: The use, reuse, or reclamation of wastes either on-site or off-site after they are generated.

- Third, consider beneficially using wastes for ENERGY RECOVERY. Some specific wastes can be beneficially used as a fuel under carefully controlled conditions to recover their energy value.
- Fourth, consider off-site TREATMENT to reduce the toxicity of hazardous wastes.
- Finally, consider LAND DISPOSAL.

NYSDOT's "Solid and Hazardous Waste Reduction Policy" outlines policies and procedures to reduce its wastes, including the following:

- *Recycling*: NYSDOT engages in agency-wide programmes to recycle such waste materials as used antifreeze and vehicle batteries.
- *Reuse*: Whenever possible, NYSDOT reuses asphalt and concrete pavements as a substitute for crushed stone in subbase and other engineering applications.
- *Waste to energy*: NYSDOT routinely collects used motor oil and compatible hydraulic fluids that are burned for fuels or space heating.

Standard operating practices that include good housekeeping are the simplest ways to reduce wastes. Other methods to reduce wastes include substituting materials; recycling and reuse; and participating in waste exchanges.

Waste Determination

The requirements for handling and disposal vary significantly for the different categories of wastes. A determination of whether a waste is a hazardous waste should be conducted for all wastes that could possibly be hazardous wastes. The definition of hazardous wastes is found in Title 40 of the U.S. Code of Federal Regulations (CFR). By definition, wastes are hazardous if they are (1) listed (specifically named, or (2) if they exhibit any of four hazardous waste characteristics 'ignitability, corrosivity, reactivity, or toxicity).

The hazardous waste determination may be conducted by using the following:

- Generator's knowledge of the waste and/or testing.

- The product's Material Safety Data Sheet (MSDS) or product label, which should indicate if an unused product would be a hazardous waste.
- Information such as ingredients, flash point, pH and disposal requirements

Testing may be required to determine whether particular wastes are hazardous. Wastes generated by Department maintenance operations which may be classified as hazardous wastes, depending on test results, include: waste paint filters, used antifreeze, used caustic solutions, waste pesticides, spent paint abrasives, waste paints, spent solvents, waste motor oil, old batteries, shop rags, waste asphalt emulsions, or waste inks.

DOTs must determine if the waste is a "listed waste" and/or has a "hazardous waste characteristic." Wastes that have certain "characteristics" (ignitability; corrosivity; reactivity; and/or toxicity) are hazardous wastes regardless of their origin. Toxic constituents can be released or leach out upon disposal, as measured by a test termed the Toxicity Characteristic Leaching Procedure (TCLP). The TCLP is an analytical test, which determines the potential of a toxic constituent (currently 40 constituents: metal, pesticide) and organic chemicals, to "leach" and become mobile and contaminate groundwater/waters upon disposal. Metals such as lead and chromium, and possibly benzene (a volatile organic) are the constituents on the TCLP list that most frequently are present in DOT wastes. Lead based paint waste, with the characteristic of toxicity, removed from bridges is one of a DOT's most frequently generated hazardous wastes.

DOTs must also consider the contaminants and/or changes to the material that could have been introduced during its use. This type of contamination may not be easily predicted by generators knowledge and may require testing of the typical waste product. Examples could include metal contamination in waste oils, degreasing solvent, or antifreeze that could be added during the vehicle operation that were not present in the virgin product. Hazardous waste generators must determine how much hazardous waste they generate and maintain records to document the amounts. Most maintenance facilities maintain Small Quantity Generator (SQG) status for Maintenance and Operations facilities by generating less than 220 pounds (100 kg) or about ½ of a 55-gallon drum, of hazardous waste in any one month, or stores less than 55-gallons of hazardous waste at any one time. If the above noted waste generation limits are exceeded, then the facility must comply with regulations for either a Small Quantity Generator Plus (SQG-Plus) or a Large Quantity Generator (LQG). State DOTs should consult with regulations and procedures particular to the state and agency.

General environmental stewardship practices include:

- Hazardous waste should be placed in the proper container/drum.
- All hazardous waste containers should be labeled with the words "Hazardous Waste" and the specific contents or words describing the waste.
- The hazardous waste container label should include the date when waste was first put in the drum and the date when the drum becomes full.
- Drums should be inspected at least weekly while stored on site.
- Arrange for off-site shipment of each full container of hazardous waste within 30 days of filling. Waste should be shipped by a permitted waste transporter with a hazardous waste manifest and disposed of at a permitted treatment/disposal facility for hazardous wastes. Waste should be removed within 90 days for LQGs and 180 days (270 days if it must be shipped more than 200 miles) for SQG.
- Maintain manifests of hazardous waste transport.
- Report spills or leaks of hazardous waste to the State Police or regulatory agency.
- DOTs should provide training at least once to all employees who handle or have responsibility for managing universal wastes on proper handling and emergency procedures and ensure that documentation of training is maintained. Training records should be

kept for a minimum of three years or for the length of employment, whichever is longer, and cover the name of the person receiving the training, the date of the training and the information covered during the training. A copy of the documentation should be sent to the DOT's central training office.

- Limitations on storing hazardous wastes at the facility's central accumulation area are dependent upon the quantity of wastes generated, with different regulations for small and large generators.
- Maintenance facilities should never exceed these time and quantity accumulation limits; otherwise the facility will be considered a Treatment, Storage, and Disposal Facility (TSDF) and be subject to extensive additional RCRA regulatory requirements.

Hazardous Waste Management Practices

Each DOT should follow guidelines and practices developed by their state and agency specialists. Environmental stewardship practices include:

Container Storage

Best practices for storage include containers:

- In good condition.
- Compatible with the wastes contained in them.
- Opened only to add or remove wastes.
- Separated from other containers holding different wastes which could cause dangerous chemical reactions.
- In compliance with container requirements for shipping wastes off-site.
- Marked with the date accumulation begins and with each subsequent date waste is placed in the container labeled with the words "hazardous wastes".
- Segregated by waste type and clearly marked to identify their contents.
- Kept within a secured area permitting access by authorized personnel only.

- Recorded in a running log of wastes accumulated in each container.
- Inspected for leaks and container deterioration weekly with the inspection results recorded.

Storage Area Practices

- Reactive or ignitable wastes located at least 50 feet from the facility property line.
- Wastes with flashpoints under 100 of located at least 60 feet from any adjoining buildings or property lines.
- Incompatible wastes segregated.
- A base and dike capable of containing leaks, spills, and accumulated rainfall.
- Adequate containment capacity necessary to hold a spill amounting to the volume of the largest container, or ten per cent of the total volume of all containers, whichever is greater, plus a reasonable amount for precipitation.
- Adequate space around containers to ensure access in the event of a spill or emergency.
- Proper emergency equipment, as needed, such as alarms, telephones, or fire extinguishers.
- A design consistent with department publication 284 titled "handbook for acquisition, development, and maintenance of the model stockpile".
- Spill and leak response measures incorporated in the PPC Plan.
- Storage of hazardous wastes in aboveground storage areas is preferred. Wastes generated by parties other than the DOT are stored at DOT maintenance facilities rights of- way unless approved by the district maintenance engineer.

Packaging

Best practices for hazardous waste packaging and transport include:

- Packages meet the USDOT or UN specifications for the wastes.

- Packages are sufficiently tight to prevent releases of materials.
- Mixing of reactive or combustible gases is prevented.
- Packages adequately closed. and
- Liquid containers should have sufficient free space above the liquid to accommodate expansion of the liquid to 130 degrees Fahrenheit.
- Product containers may be reused once for shipping wastes if the product containers:
- Are acceptable USDOT or UN specification drums for the wastes.
- Are in good condition and free of rust, damage, or leaks.
- Do not contain any incompatible residues.
- Do not carry any old marking labels that incorrectly identify the contents.
- Reused containers must be thoroughly cleaned to avoid combining incompatible wastes producing toxic vapours or explosions as well as waste mixtures that are even more dangerous than the individual wastes.
- Small items may be packaged in drums which are commonly referred to as lab packs. In general, lab packs should be packaged as noted below. The specific packaging requirements should be reviewed with the disposal contractor.
- Outside packaging should be an USDOT specification metal or fiber drum with a removable head.
- Drum construction should be compatible with the materials being packaged.
- Outside packaging should contain only one hazard class.
- Inside packaging may be one or more glass packagings not exceeding one gallon, or one or more metal or plastic packagings not exceeding five gallons.
- Inside packaging of liquid should be surrounded by compatible absorbent material capable of absorbing the total liquid contents.

Marking

All containers used to ship hazardous wastes should exhibit a USDOT hazardous waste marking completed with waterproof ink and showing:

- The proper USDOT shipping name of the waste.
- The UN or NA number.
- Generator information including name, address, and EPA ID number.
- The EPA waste number.
- The accumulation start date.
- The manifest document number.

Labeling

All containers used to ship hazardous wastes should exhibit a USDOT hazardous waste label and may require an USDOT shipping label if it meets a hazard class definition.

Manifesting

A hazardous waste manifest is a multicopy shipping document that is filled out and accompanies all hazardous waste shipments. Hazardous waste manifests are designed to track shipments from the point of generation to their final destination. All generators, except conditionally exempt small quantity generators, must use manifests to ship their hazardous waste.

- The hazardous waste manifest should be completed before shipping hazardous wastes. The manifest becomes the written record of the hazardous waste disposal. Upon shipment, forward a copy of the manifest to both the state of generation and the state of destination and retain a copy. A copy of the signed receipt manifest must be returned by the disposal facility to the generator state, disposal state and the generator.
- Copies of the manifest that are signed and returned by the treatment or disposal facility are maintained on file for five years for small quantity generators and twenty years for large quantity generators.

- If a copy of the signed manifest is not received from the waste facility within 35 calendar days, the transporter and waste facility should be contacted to determine the status of the shipment and the state/ environmental agency notified of the status.
- An exception report may be required for wastes not received after 45 days. Manifests may not be required when reclamation of the material is occurring but as a best practice the DOT may want to manifest all hazardous waste shipments because it simplifies record keeping and reporting.

Shipping

A licensed disposal contractor should provide the following services:

- Contract with a facility authorized by the EPA or the state permitting agency to treat or dispose of hazardous wastes.
- Package and label the wastes and prepare the manifest. prepare a hazardous waste characteri-zation report.
- Prepare a land band notification advising the treatment or disposal site of the standard to which the hazardous wastes should be treated. and
- Transport the hazardous wastes to the treatment or disposal facility.

Verify that the wastes were received at the waste facility by reviewing the signed manifest received from the waste facility. Liability does not end when hazardous wastes have been shipped and are no longer in the Department's possession. The Department is liable for any mismanagement of its wastes, at the current time and in the future.

Inventory and Record Keeping

Maintaining hazardous waste records is a very important part of regulatory compliance. Good record keeping proves operating compliance and may avoid problems with regulatory agencies and minimize future cleanup liabilities. Facilities judged out of compliance face legal and enforced actions, fines, and bad publicity. The following minimum records should be maintained by small quantity generators for a minimum of five years.

- Test results or waste analyses made to determine if wastes generated are hazardous.
- Monthly summaries of wastes generated which substantiate the generator category.
- This summary should indicate the final disposition of the wastes, including those not manifested.
- On-site waste accumulation records, including the date accumulation began and the quantity accumulated to date.
- In-house inspections, including deficiencies noted and when such deficiencies were resolved.
- Records of employee training.
- Generator's copies of the manifests and those returned from the destination facilities.
- Copies of land ban notifications.
- Copies of reclaiming contracts.
- Spill or leak reports.

Large quantity generators should maintain the above records and copies of quarterly reports, biennial reports, and exception reports for a minimum of twenty years.

In addition some DOTs have established Safety Coordinator positions, which with they have charged the following responsibilities:

- Ensure that an inventory of *all* hazardous chemicals is completed annually, including the product/chemical name, type of container, volume, and location of the hazardous chemicals. The Division Safety Coordinator should confirm that a current MSDS is on file for all chemicals in the inventory. Where applicable, the inventory should include Reportable Quantities for CERCLA hazardous materials and Extremely Hazardous Substances (refer to 40 CFR 302.4 and 40 CFR 355 for Reportable Quantities).
- Update inventory reports between annual inventories following any significant change in status.

- Maintain a MSDS for each hazardous chemicals used or stored in each M&O facility, and should ensure that copies of MSDS are kept at each location where chemicals are used. MSDS are to be readily available to employees, so that an employee should be able to find the appropriate MSDS within 5 minutes MSDS must contain information about the chemical including:
  - Chemical product
  - Exposure controls and company identification personal protection
  - Composition, information or ingredients properties
  - Physical and chemical
  - Hazard identification
  - Stability and reactivity
  - First aid measures
  - Toxicological information
  - Fire-fighting measures
  - Ecological 'environmental,
  - Accidental release measures information
  - Handling and storage
  - Disposal considerations
  - Transport information
- Based on the annual inventory, the Division Safety Coordinator should create a list of all hazardous chemicals that exceed 10,000 lbs., or Extremely Hazardous Substances (EHS) that exceed 500 lbs. or certain Threshold Planning Quantities (refer to 40 CFR Part 370.20 and 355). This list, hereafter referred to as the "Reportable List" should be provided to the MTS Petroleum and Hazardous Waste Management Superintendent and the Division Manager/Engineer. Petroleum products, such as diesel fuel or motor oil, are typically the only hazardous materials used by M&O in the quantities described above; 10,000 pounds equals about 1,500 gallons. M&O does not routinely use Extremely Hazardous Substances at its facilities.
- In order to comply with EPA SARA 311 and 312

reporting requirements, the Petroleum and Waste Management Superintendent should complete the following for each chemical on the Reportable List:

- Submit a copy of the state Chemical Inventory Reporting Form and MSDS for each chemical to the local fire department, the Local Emergency Planning Committee and Maine Emergency Management Agency;
- Pay chemical inventory and facility registration fees by March 1st and October 1st, respectively to the Maine Emergency Management Agency;
- Update chemical inventory reports within 90 days of a significant change in chemical inventory status.

Non-Hazardous Industrial Wastes

Some wastes, that do not meet any criteria for definition as a hazardous waste, but result from work activities are considered industrial-commercial wastes and may be disposed at municipal/commercial disposal facilities, similar to routine nonhazardous solid waste, at recycling facilities and at specialized facilities for that type of waste. Shipment, however, requires transport by permitted waste transporters, if transported in greater than exempt quantities (500 pounds/ shipment). Such waste includes used tires; non-hazardous used oil, non-hazardous waste antifreeze; other waste vehicular fluids and filters that do not meet the criteria of hazardous waste; unused products containing chemicals 'that are not hazardous wastes,; and empty drums/containers for disposal, not recycling

Permitted C&D landfills can usually accept the following types of wastes:

- Uncontaminated bricks, glass, asphaltic pavement, concrete and masonry materials.
- (Pavement containing routine intact traffic markings or that has come into contact with petroleum products through normal vehicle use of the roadway are considered clean).

- Uncontaminated soil, rock and land clearing debris.
- Wood and wood products.
- Wall coverings, plaster and drywall.
- Plumbing fixtures, electrical wiring and components containing no hazardous liquids, non-asbestos insulation, plastics that are not sealed in a manner that conceals other wastes, roofing shingles and other roof coverings.
- Empty buckets and containers (ten gallons or less) with less than one inch of residue in the bottom.
- Pipes or metal that are attached to or embedded in these waste materials.

Non-Hazardous Solid Wastes

Routine garbage, office trash, and most litter collection are considered non-hazardous wastes. Most of the adopt-a-highway trash, excluding tires and other items that are industrial or possibly hazardous wastes, are non-hazardous solid wastes. These wastes should be sent to municipal or commercial landfills or trash burning plants, and no special haulers or manifests are needed.

Specialty Waste Disposal Procedures

Specialty wastes include hazardous wastes, chemical products (including partially- used products), and/or other materials that are not disposed of by routine trash collection and require a special waste contract for disposal.

Disposing of specialty wastes is generally a two-step process:

1. Identify specialty wastes and if necessary perform laboratory testing - Known unused materials with sufficient information on their characteristics from material safety data sheets (MSDSs) or other information sources can be identified adequately for disposal. Examples include unused containers of toluene or paint with labels intact and MSDSs available. Sufficient information may also be available to identify used materials of known characteristics such

as antifreeze where the waste had previously been tested and the process generating the waste has not changed; or fluorescent bulbs which are known to be hazardous due to mercury content. For waste of unknown or uncertain identity or where contamination could be added at unknown levels to the material upon use, testing may be required to adequately identify the waste for disposal. The DOT has contracts with analytical laboratories and standard procedures for confirming suspected drum contents. The contracts with these labs are designed to characterize wastes for disposal and will meet regulatory standards without adding unnecessary testing. Call the Regional Maintenance Office or the Environmental Specialist for assistance in inventorying, identifying and testing materials for disposal.

2. Specialty waste disposal contracts - A specialty waste disposal contract can be developed for the removal and disposal of the specialty wastes as identified on the inventory. The contracts should include MSDSs and analytical results to assist the contractor in providing proper handling, record-keeping and disposal of the wastes. It is generally most cost-effective to arrange for disposal of all waste materials within a DOT Region at one time, but smaller or periodic disposal contracts may be required if storage time limits or storage space are issues. See the DOT's procedures for Storing and Handling Products and Wastes (Waste storage time limits and inspections).

Specific Guidance for Certain Waste Management Issues

NYSDOT, PennDOT, and Maine DOT guidance for specific waste management issues is included below:

Abrasives

Spent abrasives from construction projects should not be stored at maintenance facilities.

Aerosol Cans

- Disposal empty aerosol cans may be thrown in the regular trash or thrown in the metal recycle bin. Aerosol cans are empty when product can no longer be sprayed out from them. This does not include cans that have product but do not function.
- Aerosol cans that are broken, clogged or otherwise unusable must be disposed of as hazardous waste. They *must* be stored in a small container with a *lid that closes tightly and* a hazardous waste label that includes a start date and identification of the waste as aerosol cans. The container should be stored in the same area as other hazardous wastes (if there is a drum at the facility), and should be transported at the same time when regular hazardous waste pickups occur. *The small container may be any 5-gallon pail or bucket with a lid that closes.* Labels should be provided by the Petroleum/Waste Management Superintendent.
- Aerosol cans with contents no longer usable or needed must also be disposed of as hazardous waste OR the contents may be exhausted into the 30 gallon hazardous waste drum, and the empty aerosol can may be thrown into the regular trash or metal recycling bin.

Antifreeze (Coolants)

New antifreeze would not be a listed hazardous waste or fail any characteristic for hazardous waste. However, any contaminants such as chlorinated solvents, benzene, or metals that could be introduced during use must be considered to determine if the waste antifreeze could be a hazardous waste. Generator knowledge and/or representative testing of the typical waste is required to determine if it is classified as hazardous waste.

Used antifreeze should be collected in dedicated drums or tanks and clearly labeled. Disposal should preferably be by a commercial recycler who will reclaim the material.

Used Antifreeze

Used antifreeze should be collected in drums that are clearly labeled as containing "Used Antifreeze." Drums should be kept tightly closed when material is not being transferred in or out; funnels should not be kept in tank openings when not in use. Used antifreeze should be collected and recycled by the DOT.

Asbestos

Asbestos is a mineral that breaks up into very small fibers and was used for many years in making fireproofings, roofing, siding, flooring, ceiling tile and others building products. Only certified asbestos handlers can disturb, remove or package for disposal any asbestos-containing material. Any renovations or demolitions involving buildings, bridges, and utility lines that could contain asbestos must be evaluated by a certified asbestos handler. Hauling of friable (able to flake) asbestos waste requires a waste transporter permit, manifesting and waste disposal at special landfills.

(Non-friable asbestos) however, may be transported and disposed of as C&D waste. OSHA requires a visual inspection to identify materials that may contain asbestos, for future reference. If this inspection has not been performed at the facility, or if there is the possibility of finding asbestos waste along the ROW, contact the DOT environmental specialist for help and further instructions and the DOT safety guidance.

Ballasts (PCBs)

Some older fluorescent lamp (light) fixtures have ballasts that contain an oily insulating liquid that contains PCBs (Polychlorinated biphenyls) which must be disposed of as a PCB hazardous waste. PCB-free dielectric oil contained in newer ballasts can be handled and disposed of as used oil.

Assume the ballast contains PCBs unless it is marked "does not contain PCBs". The ballasts can be disposed of using specialty waste contracts or by using the Office of General Services contracts for disposal of lighting wastes from state facilities. The ballast should be separated from the lampbulb and disposed of separately. '"Fluorescent Bulbs" for bulb disposal.

Batteries

Requirements vary for batteries dependant upon their type and content and may require specialty recycling or disposal due to metal content or corrosivity. The federal Battery Act of 1996 required the phase out of mercury in alkaline batteries and required the development of recycling programmes for nickel cadmium, lead and certain other batteries. Review the information marked on the battery or provided with it, and, unless supplier information indicates otherwise, handle by the following general guidelines:

Lead Acid Batteries

- Typically vehicle batteries and small sealed batteries in electronic equipment, contain acid liquid and lead and must be recycled or disposed as hazardous waste. NYS law requires retailers/distributers to accept used automotive/truck/RV batteries back for recycling at no charge (two per month maximum without new battery purchase). Turn in the old batteries when new batteries are installed. Licensed waste transporter, manifesting of shipment, or inclusion of the battery quantities in site hazardous waste generation amounts and generator status calculations are not required.
- Nickel-Cadmium rechargeable batteries must be recycled or managed under the "Universal Waste Rule". The Rechargeable Battery Recycling Corporation 'RBRC, at 800-8- BATTERY can provide assistance in recycling; alternatively, specialty waste disposal contracts could include the recycling of these batteries in their requirements.
- Nickel Metal Hydride batteries are not specifically required to be, but should also be similarly recycled. Silver Oxide and formerly available Mercuric Oxide batteries must also be recycled or disposed of as

hazardous waste due to silver or mercury content, respectively.

- Alkaline batteries and carbon-zinc batteries are now made with no intentionally added mercury and are considered acceptable for disposal as routine municipal waste.
- Used lead acid batteries that have no cracks should be stored in a designated area, protected from the elements, with primary and emergency containment constructed of impervious material, and segregated from non-compatible materials and wastes. Used lead acid batteries should be disposed of within ninety days, but should be disposed of within one year. Lead acid batteries that have cracks are a hazardous waste.

Brush and Tree (Clearing and Grubbing) Waste

Chip and mulch, convert to compost if possible. Last choice is disposal as 'C&D Debris - Exempt C&D,". Burning is usually not allowed.

Concrete Sealers

Unused virgin concrete sealers typically have a flash point below 140° F which would classify the product as an ignitable hazardous waste. The product upon use, however, with the volatile components evaporated, would no longer meet the criteria of ignitable/flammable.

Contaminated Soil or Sediment

Contaminated soil is an industrial waste and requires disposal at municipal/commercial disposal facilities (such as sanitary landfills) reclamation facilities or at specialized facilities for the type of contamination present.

The potential for the contaminated soil to be a hazardous waste due to characteristics such as flammability or toxic metal content must also be considered. If soil or sediment contamination is suspected, call the RLA/ESU to help arrange for further investigation and possible testing. Soil or sediment may be contaminated if it is discolored or stained, or smells like fuel or sewage.

Culvert and Catch Basin Cleaning

Uncontaminated grit and sediment from culverts and catch basins is normally disposed of as C&D waste and is not considered contaminated unless it smells like petroleum, fuel, or solvents, or is mixed in with other wastes like roadside trash.

Drums and Containers

Drums and containers that have had all of the contents removed by common practices and have less than 1 inch and less than 3 per cent of the original product are considered "empty" and nonhazardous, even if the material they contained 'such as solvents or coatings with flashpoints below 140°F, would otherwise be classified as a hazardous waste. "Empty" containers may be returned to the manufacturer or sent to a reconditioner or handled as scrap metal, cardboard, etc. and are exempt from waste transporter requirements when destined for such reuse. "Empty" containers are nonhazardous industrial wastes when otherwise disposed. Small containers of up to ten gallon capacity are, however, considered C&D debris and can be disposed of as such.

The original product label and hazard warnings must be left on drums or containers until they are empty as described above and no longer pose the indicated hazard. Remove or obliterate the label and mark the drum "empty" as soon as the drum is empty by these criteria. The hazard markings must be removed from an empty drum meeting these criteria prior to removing from the facility. Drums are considered empty when there is less than one inch of product remaining and less than three per cent of the original product in the drums. Empty drums should be stored neatly in a designated area with lids and bungs secured with end drums blocked to prevent rolling. The empty drum storage area should be in a location that does not permit surface water to collect or wash through the storage area. Empty drums should be disposed of at least annually. Abandoned drums or containers of unknown substances that are found along the ROW are handled similarly to spills of hazardous substances on the ROW and may need to be reviewed by police. Further details are available in the following discussion on drum management within facilities.

Fills

Suitable fills are environmentally inert, uncontaminated, non-water soluble, solid materials. Only suitable fills should be used to level an area or bring it to grade provided the area is not located in a wetland.

Suitable fills may be commingled with other suitable fills at maintenance stockpiles while being stored prior to placement in a fill area.

Examples of suitable fills include:

- Shoulder cuttings
- Bituminous asphalt excavations pipe excavations 'but not metal or plastic pipes,
- Crushed Portland cement concrete without exposed reinforcement bars
- Bricks and solid masonry blocks
- Clearing and grubbing vegetation

Fluorescent Bulbs

Typical spent fluorescent bulbs (lamps) are hazardous wastes due to mercury content. Intact (not crushed or broken) fluorescent lamps are eligible to be handled as "universal wastes" allowing for somewhat reduced regulation. Some manufacturers are marketing lamps with lower mercury content; these lamps may not be hazardous wastes when spent. Unless the lamps are marked (or otherwise identified) as low mercury content lamps, assume that the lamps must be handled and disposed of as a universal or hazardous waste.

The waste bulbs can be disposed of using specialty waste contracts or by using the Office of General Services contracts for disposal of lighting wastes from state facilities. Lamps that are marked or identified as low mercury must be evaluated to determine if they are a hazardous waste; manufacturer's data may be used to support a determination that particular lamps are not a hazardous waste. Note: The ballast should be segregated from the lamp and may also be a hazardous waste due to PCB content.

Fuel Filters

Used gasoline or diesel fuel filters are classified as hazardous wastes because they typically have the characteristic of ignitability or toxicity for benzene. These should be stored in closed containers, separate from other wastes, and labeled, handled and disposed as hazardous wastes. However, if the fuel filters can be drained of all free liquids, they can qualify as scrap metal and be recycled at a scrap metal facility, under the scrap metal exemption.

Grease and Tar

Collect grease and soft tar in separate containers with proper labeling. Include these containers for disposal by a specialty waste disposal contract. Greasy rags should be collected along with oily rags and absorbents and properly disposed by arrangement with the Petroleum/Waste Management Superintendent.

Greasy rags should not be disposed of in the regular trash. Any waste grease should be disposed of as special waste by arrangement with the Petroleum/Waste Management Superintendent.

Hazardous Substances in Equipment

Some equipment contains hazardous substances that may require special handling and disposal. Examples include switches or thermometers that contain mercury, or ballasts and light fixtures with PCBs. Call the environmental specialist or landscape architect with specific questions.

Hydraulic Fluid

Hydraulic fluid products such as brake fluid, transmission fluid and power steering fluid are chemically different from motor oil, but NYSDEC Used Oil regulations considers them used oil and allows mixture and recycling along with used oil. The recycler/disposal firm should be consulted, however, on their specific requirements. The fluids also must not be contaminated with any solvents or other materials that could cause them to be hazardous wastes.

Medical Waste (Used Syringes or Needles)

Used hypodermic needles and syringes are sometimes discarded at rest areas or along ROW. They are classified as household waste, not as "regulated medical waste" as defined in the Public Health Law when they are found at public recreation spots such as rest areas. Used needles and other "sharps" can poke workers, and some bloodborne diseases like hepatitis can be transmitted if the virus is present. by the Petroleum/Waste Management Superintendent.

(The AIDS virus is unlikely to live more than an hour outside a human host, but should also be considered a risk.) Carefully place the syringe in a container and label with a biohazard sign (or use red containers). Disposal should be at a local hospital or other facility that can accept medical waste.

Oil

Environmental stewardship practices for used oil include the following:

- Collection: Used oil produced at M&O facilities should be collected for recycling or burning in approved used oil furnaces. The DOT should collect used oil in drums or tanks clearly marked "USED OIL" for transportation to locations with used oil burners. Care should be taken not to contaminate the used oil with hazardous materials, or other non-approved materials.
- Storage: Oil storage tanks or drums should be located in areas with an impervious floor 'such as concrete. Tanks should have secondary containment where risk of damage is high, or where the impacts of a spill would be severe. Metal tanks or drums should not be

in direct contact with the ground (contact with a dry floor is permitted); only tanks with double-walls and a leak detection system maybe placed on or below the ground surface.

- Used oil tanks or drums should be clearly labeled as such, and should be posted with a list of what is permitted to be dumped into the tank. All used oil tanks that could be accessible by the public 'outside, should be locked to prevent unauthorized dumping of hazardous or non-approved materials into the tanks.
- Storage tanks should be inspected by the Petroleum/ Waste Management Superintendent and/or Division personnel at least annually for structural integrity. The supervisor should keep a record of the inspection at the facility. The DOT should arrange for recoating or replacement of tanks that are rusting and/or pitted. Existing spill containment structures should be inspected annually for structural integrity and repaired if necessary (such as filling in cracks in concrete).
- Suspected contamination of oil waste: If the used oil is suspected of containing contaminants, such as solvents, brake cleaner, or toxic metals, then the contaminated oil should be contained separately and disposed of as hazardous waste.
- Used oil filters are considered a non-hazardous waste if the used oil is removed from the filter. The filter may then be preferentially recycled as scrap metal or otherwise disposed of as non-hazardous waste. Environmental stewardship practices for used oil include the following:
  - Oil filter recycling should be available at all M&O facilities.
  - Collection procedures should be established for each location.
  - The drained oil should be combined with other used oil from the site for recycling.
  - Properly drained or crushed filters should be

placed in a container marked "used oil filters" and disposed of every ninety days.

In order for the oil to be considered removed, filters should be gravity hot-drained by one of the following methods:

- Puncturing the filter and hot draining. EPA recommends that hot draining occur at or near engine operating temperature for at least 12 hours.
- Hot draining and then crushing the filter.
- Dismantling and hot-draining.
- Any other equivalent hot draining method that will remove used oil.

Environmental stewardship practices for use of oil absorbents include the following:

- *Types and use of absorbents:* Maintenance workers should be provided with, and trained in the use of, spill pans and absorbents to minimize spills or drips on shop floors, and to reduce the quantity of absorbents used. Absorbent pads and pans should be provided as appropriate for the following:
  - Under spigots on oil dispensing drums and tanks;
  - Under leaky hydraulic valve boxes and hoses (hoses should also be capped and elevated where possible;)
  - Around used oil collection tanks and associated filter pipes and drainage racks;
  - Under vehicles and machinery undergoing repair or maintenance.
- *Collection of used absorbents:* Oily absorbents (including Speedi-Dry) should be collected in a sealable container (such as a drum) and properly disposed by arrangement with the Petroleum/Waste Management Superintendent. Oily waste should not be disposed of with regular trash.
- *Substitute absorbents:* When possible, M&O should use substitute absorbent materials which are not clay-based. DOTs should maintain an approved list of absorbent materials.

Paint

Most unused paints, including waterborne, have a flashpoint below 140 F and therefore require handling and disposal as an ignitable waste. If the paint contains lead or chromium, the potential for the waste to have a toxicity characteristic for lead or chromium must also be considered. Yellow waterborne traffic marking paint may contain significant amounts of lead chromate. Limited testing of the waterborne yellow paint, however, has indicated that it did not fail the toxicity characteristic tests. Consult the Material Safety Data Sheet (MSDS) and the RLA/ESU for further information.

Dried Paint Chips and Flakes

Dried pavement marking paints and other non-lead dried paint are non-hazardous industrial wastes, requiring disposal at a municipal landfill. These dried paints would include markings purposely removed/milled from the road surface, but would not include the paint markings incidentally present on an entire removed section of roadway which would qualify as C&D debris. Some dried paints containing lead such as former lead- based bridge paint removal debris are considered a hazardous waste for their lead content and must be disposed of as hazardous waste. Testing of typical dried yellow pavement marking paints (waterborne and epoxy), however, has indicated that, although lead and chromium are present in significant concentrations, they are under regulatory levels for hazardous waste. Note: Landfills, however, may be unwilling to accept paint waste and/or may require additional testing. Dried paint wastes may also be collected, stored and disposed of by the specialty waste disposal contract.

Paint Thinner

Most paint thinners are organic solvents that would be listed or ignitable wastes. Store and handle these as hazardous wastes.

Parts Washer Wastes

Spent solvents from parts washers may be hazardous

wastes due to ignitability. The solvent used in Safety Clean parts washers typically has a flashpoint below 140°F and would be an ignitable waste upon disposal. Any contaminants such as metals that could be added through use must also be considered. Typical spent filters (bag and cartridge filters) from Zep parts washers have been tested for contaminants including metals that could be introduced during its operation and were under regulatory limits and are therefore determined to be a non-hazardous waste.

Parts Cleaner Waste

- *Parts cleaner solvent:* All M&O facilities should use parts cleaner solvent, approved by the Petroleum/Waste Maintenance Superintendent, that would not be classified as a hazardous waste if contamination by metals or other chemicals did not occur during use. For example, do not use solvents that contain chlorinated compounds or have a flash F° point less than 140. Maintenance personnel should not contaminate the parts cleaner with hazardous materials, such as chlorinated solvents often contained in spray cleaners.
- *Disposal of used parts cleaner solvent:* Spent parts cleaner and parts cleaner filters will be disposed of as hazardous waste 'note: used solvent may be hazardous even if the virgin product is not hazardous, due to contamination during use., M&O *Hazardous Waste Management Procedure.*
- *Registration of parts cleaners:* The DOT Safety Coordinator should register any parts cleaners containing volatile solvents with the state environmental agency. At a minimum, the DOT should keep an inventory of parts cleaners in service, and will notify the environmental agency of the size or volume and type of parts cleaner, and the type of solvent used. The DOT should determine if additional requirements apply depending on the volatile organic content of the solvent.

Pesticides 'Includes Herbicides and Insecticides

Keep all pesticides in their original, labeled containers, and keep its Material Safety Data Sheet (MSDS) on file at the facility. Partly used containers should be saved for next use. Pesticides that can not be used must be disposed of by a specialty waste contract.

Empty containers of non-combustible pesticides may be disposed of as non-hazardous waste after triple rinsing (with the rinse water used to make up the next batch of herbicide); or, for ready-to-use (do not require dilution) pesticides, after draining for one thirty second period.

Petroleum Contaminated Soil

Soil materials contaminated with petroleum products, including (but is not limited to) gasoline, heating oils, diesel fuel, kerosene, jet fuel, lubricating oils, motor oils, greases, and other fractions of crude oil are considered petroleum contaminated soil and require disposal as industrial waste.

Refrigerants

Refrigerants such as Freon are used in air conditioning systems and contain chlorofluorocarbons (CFCs) which pollute the air. Freon and other refrigerants must be removed and recycled by workers with EPA-approved training. Maintain records that show the name of the recycling facility that removed the refrigerants.

Shop Rags (or Shop Towels)

When rags are used to clean up known nonhazardous waste materials such as non-hazardous cleaning solvents or hydraulic fluid or motor oil, the rags would not be a hazardous waste. If, however, rags were used to soak up a material that would be a hazardous waste (toluene or chlorinated solvents for example), the rag could be a hazardous waste.

Hazardous waste rags are not regulated as hazardous wastes if they are sent out to be cleaned and returned for re-use. All used rags, shop towels, and clothing soiled with parts cleaner, gasoline or diesel fuel, used oil, etc. Should be stored and managed in fire-proof or fire-resistant containers and must not be so saturated that they can drip any free liquid.

Since DOTs should avoid using chlorinated solvents or hazardous waste cleaning solvent or other listed materials, rags for disposal should not be non-hazardous industrial wastes. Any rags, however, that were used for potentially hazardous waste materials could be hazardous wastes requiring disposal as specialty hazardous wastes and should be kept separate from non-hazardous waste rags. Contact the DOT environmental specialist for assistance in determining if non-routine rags for disposal are hazardous wastes.

Sorbents (Speedi-Dry or Sorbent Pads)

When sorbents are used to clean up spills from known nonhazardous sources such as hydraulic fluid or non-flammable (non-chlorinated) parts washers, the used sorbent is not a hazardous waste and may be disposed of as routine nonhazardous waste.

Sorbents used to clean up known hazardous wastes, however, would also be a hazardous waste. When sorbents are used to clean up spills from unknown sources, they could be hazardous wastes and should be tested. Call the RLA/ESU to arrange for testing and/or disposal as specialty wastes.

Street Sweepings (Shoulder Maintenance)? Routine street sweepings are not considered contaminated and can be handled like fill or sent to a C&D (construction and demolition) or municipal waste landfill.

Street sweepings should be handled as contaminated soil if they smell like petroleum or solvents, or contain considerable roadside litter such as paper, cigarette butts, plastic, etc. Contaminated street sweepings must be sent to a municipal landfill. This topic is covered in greater detail in roadside, non-vegetative management practices.

Tires

Waste tires and scrap tires collected along state highways

can be stored for up to 18 months. A permit is often required to store more than 1,000 tires. Waste tires can be sent to a landfill, recycler, or trash-burning incinerator, but some landfills do not accept scrap tires because they are bulky and tend to "float" to the top of the waste pile. Some cement kilns or burn plants can burn tires for fuel. Check with the local waste hauler or landfill to see how to dispose of waste tires in the area. Used tires should be stacked by size, type of construction and vehicle use and stored under roof or tarped. Used tires should be disposed of annually or whenever they number 500 in count, whichever comes first.

Routine trash and Adopt-A-Highway Waste is considered non-hazardous waste and can be disposed of as routine refuse at municipal/commercial landfills or disposal facilities. Some wastes need special handling. Such wastes include Abandoned Drums and Containers, Medical Waste, Tires, etc. See pavement and materials recycling section.

Universal Wastes

Certain common hazardous wastes that were considered to be low risk have been designated as "universal wastes" with somewhat reduced regulation. Universal wastes currently include spent batteries, certain unused pesticides, fluorescent bulbs containing mercury and mercury thermostats. Manifests are not required for shipment 'although permitted waste transporters are required for transport of >500 pounds/shipment, and wastes may be accumulated on site for up to 1 year. Small quantity handlers (up to 5,000 kg at one time) do not need an EPA ID number.

Used Oil (Waste Oil) ? Used oil that is destined for recycling or burning for energy recovery is not regulated as a hazardous waste. Examples of used oils include spent motor oil, hydraulic oil, cutting oil, transmission fluid, fuel oils, gear oil and greases (*Note*: waste fuel oil is not regulated as used oil). Used oil should be collected into clearly labeled tanks or drums. Do not mix any other materials such as solvents, antifreeze or gasoline with the used oil.

If some type of hazardous waste such as solvents, degreasers, etc. are mixed with the used oil, then the entire volume may be classified as a hazardous waste. The used oil should be sent to an authorized recycler or fuel blender using a permitted Waste Transporter.

DRUM/CONTAINER MANAGEMENT

Drum or container management applies to any used metal, plastic, fiberglass or laminate drums/containers, typically 5 gallons in capacity or larger, used for bulk storage of liquids, solids or waste materials.

The Maine Department of Transportation developed the following procedure and environmental stewardship practices for managing drums and containers:

- Establish a procedure for managing drums and smaller containers (such as 5-gallon pails) on Maintenance and Operations facilities. Procedures should be audited (checked for compliance and improved) annually.
- Ensure personnel are trained to properly manage used drums/containers to prevent releases of drum/container residues to the environment and to maintain a safe, neat working environment. All employees who use or manage drums/containers should be trained at least annually and easy-to-follow guidelines for proper management of used drums/containers should be prepared. Training records should be kept in a central location.
- Routine inspections of facilities by the Division Manager/Engineer, Division Safety Coordinator, or the Environmental office should include a review of drum/container management practices.
- Crew supervisors should have primary day-to-day responsibility for compliance with drum management policies and procedures.

Drum/Container labeling

- All drums/containers should either be labeled "empty"

or with the intended contents, such as "trash" or "oily rags."

- Empty drum/containers need not be labeled if they are neatly stockpiled in a designated area marked with a sign "Empty drums".
- Labels may be adhesive type (waterproof if exposed to the weather), or painted. Painted labels should be legible. Old labels should be removed or completely covered.
- Drums containing inert metal parts do not need to be labeled.

Drum/Container Purchasing

The DOT should not purchase or obtain drums/containers from any source other than a vendor supplying new or properly reconditioned drums; an approved vendor.

Recycling, Reuse, and Disposal of used Drums/Containers

- Only sound drums/containers in good condition, free of substantial rust, cracks or dents, should be used by maintenance and operations
- Drums/containers should be thoroughly drained so that no "flowable" product remains. Product should be used for its intended purpose, or, if obsolete, disposed properly based on the waste type.
- Only drums/containers that formerly contained the following should be reused:
  - Motor oil
  - Hydraulic oil
  - Gear oil
  - Transmission fluid
  - Grease
  - Antifreeze
  - Soap/detergent

Drums containing other materials should only be reused if approved by the Division Manager/Engineer in consultation with the Environmental Office (ENV) and Office of Human Resources, Safety Section, if appropriate.

- Prior to reuse, drums/containers should be cleaned by wiping or another approved method that does not result in the release of drum/container residuals to the environment (soils) groundwater or surface water.
  - Cloths or other materials used for wiping should be cleaned and reused or disposed of properly. For example, oily rags should be disposed in the facility oily waste container.
  - Alternatively, arrangements can be made with an outside vendor for reconditioning of used drums.
- Empty drums/containers, containing no residual product but not suitable for DOT reuse should be recycled for scrap metal as part of a metals recycling programme.
- Drums/containers that contain residues of oil, chemicals or other waste should be disposed properly in accordance with the waste type.

SPILLS AND LEAKS

Numerous federal and state regulations specify extensive requirements for the prevention of spills and leaks of hazardous wastes at DOT facilities. In addition, many federal, state, and local agencies should be immediately notified of a hazardous waste release.

Severe penalties and fines are often imposed for failure to notify. The first and most important step in a spill or leak response is to safely contain the spill and stabilize the situation by following the methods described in the facility's PPC Plan, and then to notify the proper authorities.

Spill prevention and control procedures and practices are typically implemented to prevent and control spills in a manner that minimizes or prevents the discharge of spilled material to drainage systems or watercourses. Spill prevention and control procedures are typically implemented wherever chemicals and/ or hazardous substances are stored. Substances may include, but are not limited to, soil stabilizers, dust palliatives, herbicides, growth inhibitors, fertilizers, de-icing chemicals, fuels, lubricants and other petroleum distillates. To the extent that the work can be accomplished safely, spills of oil, petroleum products, substances listed under Title 40 of the Code of Federal Regulations (CFR) Parts 110, 117, and 302, and sanitary and septic wastes should be contained and cleaned up immediately.

Caltrans and NYSDOT employ the following spill prevention practices:

- To the extent that this action does not compromise cleanup activities, spills should be covered and protected from stormwater run-on during rainfall.
- Spills shall not be buried or washed with water.
- Used cleanup materials, contaminated materials and recovered spill material that is no longer suitable for its intended purpose should be stored and disposed of in conformance with these special provisions.
- Water used for cleaning and decontamination shall not be allowed to enter storm drains or watercourses.
- Water overflow or minor water spillage should be contained and shall not be allowed to discharge into drainage facilities or water courses.
- Proper storage, cleanup and spill reporting instructions for hazardous materials stored or used on a project site should be posted at all times in an open, conspicuous and accessible location.
- Waste storage areas should be kept clean, well organized and equipped with ample cleanup supplies that are appropriate for the materials being stored. Perimeter controls, containment structures, covers, and liners should be repaired or replaced as needed to maintain proper function.
- Spill control cleanup materials should be located near material storage, unloading, and use areas.
- Update spill prevention and control plans and stock appropriate cleanup materials whenever changes occur in the types of chemicals stored on-site.

- Inform and remove unnecessary employees from the area.
- Determine the identity and hazards of the material and any personal protective equipment such as impermeable gloves required for handling.
- If the spilled material is flammable, remove any open flames or sources of ignition. Use non-sparking tools and grounding wires if needed.
- Stop additional material from spilling at its source if possible. For example, plug a leaking hole in a barrel or turn the barrel so that hole is on top.
- Plug any drains that may be impacted.
- Contain the spill by placing absorbent "socks" or sand to prevent the spill from running into storm drains, bare soil, large surface areas, etc.
- Pump large quantities to an empty drum that will hold the material.
- Collect smaller quantities and/or remaining liquid by absorbing liquid with absorbents or sand. Gently scoop or sweep up the residue and place in empty container.
- Label all containers of spill collection and debris as soon as possible.
- Do not try to clean up spills of unfamiliar materials if adequate hazard communication information is not in place.

Emergency Preparedness and Response Planning

Emergency preparedness and response plans are typically required by law, regulation, or DOT policy for each facility/depot.

- Each emergency preparedness and response plan should show the issue and revision date, a list of holders of the copies and their locations, and a log of revisions issued.
- Division Manager/Engineers should establish a schedule to test the effectiveness of the emergency preparedness and response plans, using drills or mock

emergencies. At least one test or drill should be conducted annually. (A real emergency may be considered as a test).

- Division Managers/Engineers should hold a debrief session after each real emergency, test or drill, and prepare a written summary of lessons learned and any necessary revisions to facility operations, or to the plan.
- Emergency plans should be reviewed by the Division Manager/Engineer in consultation with environmental staff at least once per year, or when significant changes are made in operations or facilities, to determine if the plans require revision. Revisions should be distributed to holders of all copies, and revision dates will be noted in the revision logs.

Communication and Training for Emergency Preparedness and Response

- DOTs should provide for emergency preparedness and response training for new employees, and annual refresher training for current employees, based on the current emergency preparedness and response plans. Major revisions of a facility plan will require training updates in a timely fashion.
- Current copies of each emergency preparedness and response plan should be provided to crew supervisors, to appropriate local emergency response agencies 'fire, rescue, police, local emergency management agency, etc., and to appropriate state and federal agencies as required.
- A copy of the plan's, should be kept at strategic locations in the facility so it can be easily accessed in an emergency.
- A current copy of each emergency preparedness and response plan should be provided to major contractors who will be operating on site for an extended period of time, as well as emergency preparedness and response training, as appropriate. Contractors working

on site may be required to provide a copy of their own emergency preparedness and response plan for their own work. a spill prevention plan for equipment operated on site.

Spill Prevention Control and Countermeasures (SPCC) Plans

Missouri DOT identified the following information from a U.S. EPA teleconference to help aid facilities in understanding and working with Spill Prevention Control and Countermeasures Plans and new requirements.

What is to be Covered in the Plan?

- Oil of any kind and hazardous materials 'list found at,
- Total volume of materials listed, total over 1,320 gallons counting containers 55 gallons or over
- Material can reach Waters of the U.S., which include lakes, rivers, streams, dry creek beds, ditches, wetlands, and tributaries to these.
- Manmade structures, dikes, equipment are not considered reasons that our oil cannot leave our property and cannot be used as a reason to not have a plan in place

Requirements for Preparation and Implementation

- Professional Engineer (PE) must certify
  - They are familiar with the rule
  - The PE or agent has visited and examined the facility
  - The plan is prepared in accordance with good engineering practice (considering applicable industry standards) and with rule
  - Testing and inspection procedures are established
  - The plan is adequate for the facility
- The plan
  - It must be kept at the nearest manned facility
  - Provided to the inspector during normal working hours
- Written report must be submitted

  - Spill >1,000 gallons to the environment
  - Two reportable spills of >42 gallons in a year
  - Submitted to EPA and MDNR
  - EPA and MDNR may require amendments to the plan at that time
- Plan amendments by owner/operator are required
  - When there is a physical change affecting potential for a spill (such as taking down or adding tanks)
  - Review plan every five (5) years
  - Document review and amend to include more effective spill control technology
  - Only technical amendments must be certified by a PE
  - Changes in phone numbers, supervisor, employees and other non-technical changes do not need a PE re-certification
- Signed by owner/operator (supervisor or superintendent)
- Follow the sequence rule 112.7 for the plan
- List equivalent environmental protection
- Provide detailed diagram of facility
- Describe prevention and countermeasures
  - Type of product and capacity of each container
  - Prevention measures for handling and storage
  - Discharge and drainage controls
  - Countermeasures, disposal and reporting of discharge
- Spill prediction section on what would cause a spill and where would it flow
- List and describe containment
  - Dikes or berms that are sufficient impervious to contain spilled material until it is cleaned up
  - Curbing, culverting, gutters or other drainage
  - Weirs, booms or other barriers
  - Spill diversion or retention ponds
  - Show direction drainage for all containers
- If a facility cannot physically put in containment

  - Explain why
  - Conduct integrity testing of tanks and leak testing of pipes and valves more often
  - Develop a contingency plan (response plan) (40CFR109)
  - Have a written commitment of manpower and equipment to stop a spill and clean it up (good idea for all our plans anyway)
- Records must be kept for three years with plan and signed by supervisor, and include the following:
  - Frequency of testing and inspection of tanks
  - For tank, piping, valve inspection and testing
  - Water drained from containment
  - SPCC plan review every five (5) years
- Training and personnel requirements
  - Annual or as determined by DOT
  - MoDOT conduct training for employees on equipment and spill prevention and response procedures
  - One person must be designated as responsible for SPCC requirements
  - Conduct and document periodic briefing on recent problems and new spill prevention measures 'frequency determined by need,
- Security
  - Facility or area with tanks must be fenced unless occupied 24 hours a day to deter vandals, commensurate to location, and gated and locked when not occupied
  - Master flow and drain valves on tanks and containment must be secured in the closed position unless in use (locked)
  - Pump starter controls must be locked and only accessible to authorized personnel
  - Loading and unloading connections must be capped when not in service and locked if they can result in a release

  - Adequate lighting to detect and cleanup spills at night and deter vandals
- Loading and unloading areas
  - Must have secondary containment for largest tank plus 10 per cent such as:
  - Quick drainage system
  - Catch basin or treatment system
  - Curbing or speed bump type berms
  - Diversion to secondary containment
  - Trenches, sumps, USTs etc.
  - Seal drains
  - Cleaned up after spill
- The following can be made the responsibility of the supplier
- Monitoring continuously during filling or loading operations
- System to prevent trucks from leaving prematurely
- Physical barrier
- Warning light
- Wheel chocks
- Or vehicle brake interlock system
- Inspect vehicles for leaks before departing
- Field constructed tanks. Evaluate brittle fracture and take appropriate action if the container undergoes the following
- Repair
- Alteration
- Reconstruction
- Change in service
- Conformance to oil pollution prevention rules
- Discuss conformance with the rules in the SPCC plan
- Discuss conformance with any applicable more stringent state rules, regulations and guidelines
- Explain where and why there is non-compliance with any of the rules

Facility Drainage

- Secondary containment must
  - Hold the contents of the largest tank
  - Plus sufficient freeboard for 25 year even rainfall (10 per cent)
  - Be sufficiently impervious to hold a spill until it can be detected and cleaned up
  - Free of vegetation that would compromise the containment or inhibit inspections
  - Water drained from diked areas to allow room for the largest tank's contents 'always remove oil from water first,
- Restrain drainage
  - By manual controlled pumps
  - Manual siphons
  - Discharge to approved treatment system
  - Manual valves
  - May not use flapper-type drain valves
    a. Catch basins
    b. Underflow designed piping
    c. Through wall valved pipe
    d. Retain oil until discovered
    e. Located outside flood prone areas
  - Engineered flow control with curbing, trenches, dikes, terrace, or diversion pond
  - Minimum of two lift pumps at lift stations for treatment systems
- Empty containment
  - Uncontaminated storm water
  - Only after visual inspection
  - Removal of oil from water
  - Only when safe to discharge to approved treatment system
  - Remove all trash and properly dispose of material
  - Clean any oil from containment
  - Record each event with Date, Time, Volume

released (may need to estimate), and Where discharged

Bulk Storage Containers

- Must be compatible with material stored inside
- Contain necessary pressure and temperature
- Design so containment can be provided
- Keep any bypass valves sealed and closed (locked)
- Record inspection and tank testing
  - Visual inspection monthly of tank and secondary containment, looking for:
    a. Corrosion and deterioration
    b. Product discharges inside containment
    c. Dents or holes
    d. Scraped paint
    e. Cracked valves or fittings
    f. Any other visual damage the may jeopardize the integrity of the tank or containment
    g. Cracks
    h. Facility effluents
    i. Testing by certified tester or equivalent
    j. Performed according to manufactures standards and recommendations
    k. Conducted at intervals recommended by industry or engineer (above ground tanks are usually tested every 10 to 15 years)
    l. Must be done when changes are made that may affect the integrity of the tank or it was damaged.
  - Intervals must be in plan along with records of inspections and testing
- Inspect water before discharge under responsible supervision
- Test each aboveground container for integrity on a regular schedule, and whenever you make material

repairs and keep a record. Methods include but are not limited to

- Hydrostatic testing
- Radiographic testing
- Ultrasonic testing
- Acoustic emissions testing
- Other non-destructive shell testing

Transfer Operations, Piping and Pumping

- Buried metal piping must be coated, wrapped and cathodically protected
- Exposed buried line must be inspected for deterioration and repaired as needed
- Out of service piping must be drained, labeled and capped
- Pipe supports must be designed to minimize corrosion
- Above ground piping and valves must be inspected regularly and recorded
- Buried pipes must be leak tested when installed, modified, construction, relocation, or replacement
- Pipes must be protected from vehicular traffic with warning signs and barriers

Records to Maintain

- Emergency contacts
- Substantial Harm Criteria Checklist
- Maps showing tanks, piping, loading areas, and where spill would flow if they got outside secondary containment
- Secondary containment calculations
- Spill notification form
- Other records
  - Employee sign familiarity sign sheet
  - Employee training roster
  - Inspection records for tanks, valves, and piping 'monthly visual inspections,

Finding and Resolving Illicit Connections, Illicit Discharge, and Illegal Dumping

Illegal dumping and spills can encroach on DOT properties. Accessible DOT properties, such as safety roadside rest areas, vista points, turnouts, and weigh stations, provide tempting places for illegal dumping.

Illegally dumped and spilled materials have potential to flow into receiving waters. This section reviews environmental stewardship practices for minimizing the impact of illegal dumping and spills that are outlined in the Caltrans Stormwater Quality.

- Look for potential dumping at drain inlets, open channels, and municipal storm drain system tie-ins. Warning signs can include:
- Visible signs of staining or unusual colours to the pavement or surrounding adjacent soils.
- Pungent odours coming from the drainage system.
- Discoloration or oily substances in the water or stains and residues detained within ditches, channels or drain boxes.
- Dumping of debris or medical waste at a particular location, where the proximity to the drain system could impact water quality.
- When any field maintenance employee witnesses or discovers a suspected illegal dumping to a DOT storm drain system, it should be report it to the supervisor.
  - If the substance is known to be hazardous, suspected of being hazardous, or cannot be identified, notify the District Maintenance HazMat Manager immediately.
  - If an illegally dumped substance within DOT right-of-way has the potential of entering a municipal drain system, notify the supervisor and the District Maintenance Stormwater Coordinator so that the downstream municipality can be contacted.
- Provide protection for adjacent drain inlets to prevent

entry of the illegally dumped substance, if it is safe to do so.

- Perform cleanup and corrective actions for illegal dumping and spills on DOT right-of-way in accordance with DOT procedures and District *HazMat Spill Contingency Plan.*
- Maintenance supervisor should follow up on the incident to ensure the appropriate agencies have been contacted and corrective actions have taken place.

DOT Contribution to Local Spill Prevention Initiatives

In some cases, DOT maintenance staff have taken the initiative to contribute to local environmental protection and conservation initiatives. The Skaeneateles Lake Watershed provides drinking water to a quarter million people and is completely encircled by highways.

NYSDOT's Region 3 Maintenance Group formed a partnership with the City of Syracuse Board of Water Supply, NYS Department of Environmental Conservation (NYSDEC) and local officials and took the lead in addressing spill-related water quality concerns.

NYSDOT identified 15 sites and potential contaminant pathways that most directly threaten water quality and mitigated those sites through the installation of stone check dams/retention basins to treat daily run-off and to temporarily retain any spills that may occur at these sites. Also, all Department maintenance staff working in the 20-square-mile watershed have been provided sensitivity training and spill clean-up materials. In addition, local DPWs in the watershed have been provided with similar spill control materials and the local fire department has been provided with culvert plugs, absorbent materials and a commercial grade spill containment boom.

VENTILATION AND EXHAUST SYSTEMS AT MAINTENANCE FACILITIES

Some operations and process ventilation that can release

contaminants to the air may require an air permit from a state regulatory agency.

For example, paint spray booths are likely to require an air quality permit if:

- More than 25 gallons of paint and solvents 'combined, are used in a month,
- The paint spray booth is located where air quality is poor for ozone, or
- Exhaust gases from sanding and painting do not pass through filters or other emission control devices.

FACILITY INSPECTION AND REINSPECTION TO ACHIEVE ENVIRONMENTAL

Facility inspection and reinspection programmes are discussed in detail, Organizational Environmental Stewardship Practices. Maine DOT and Mass Highway boast excellent examples of facilities auditing procedures and practices. Caltrans' reinspection programme has found that follow up on initial inspection results has improved housekeeping, spill cleanup, hazardous materials and waste storage and documentation practices.

Specific criteria were defined for selecting facilities for reinspection:

- Number of compliance action items identified from the initial inspection
- Progress made on action items based on documentation submitted by the facility
- Size of the yard and number of crews
- Proximity to the nearest downstream watercourse
- Types of materials stored at the facility

Caltrans identified the following reasons for follow up inspections of maintenance facilities:

- Enhance compliance at facilities considered to be facing the greatest stormwater management challenges.
- Provide facility-specific training to improve staff understanding of stormwater requirements and Environmental stewardship practices.

- Prepare facilities for regulatory agency inspections.
- Review and offer assistance in filing procedures for stormwater documentation.

DOTs that employ such audits and inspections, often do so on a rotating five year basis, so 20 per cent of facilities are audited or reinspected in any given year.

YARD AND FLOOR DRAIN MANAGEMENT

Many DOT facilities have floor drains in buildings and surface drains in the "yard" that discharge washwater and stormwater through underground pipes or open ditches, a ditch, or a right-of-way culvert, or off-site to a neighbouring property and, potentially, to the waters of the state.

Shop floor drain effluent can include motor vehicle fluids spilled as a result of vehicle maintenance and repair, and washbay washwater, especially from salt truck and salt bed washing during snow and ice season.

Flows off of maintenance yards may ultimately reach waters of the state, either indirectly or via direct discharge into a side ditch. Discharges from floor drains to surface and groundwaters are generally regulated through the National Pollutant Discharge Elimination System (NPDES, and state administered NPDES programmes) often called SPDES. NPDES/SPDES permits are issued and required for discharges to waters of the state. Some facilities have installed oil/water separators as interceptors; however, such separators are not effective in removing soluble contaminants such as salt from the discharge.

PROCEDURAL PRACTICES AND OTHER NON-STRUCTURAL BMPS

Stormwater/Washwater BMPs for Surface Run-off

Much of the impact from surface stormwater/meltwater run-off can be significantly reduced by removing contaminants from the path of the sheet flow and point discharges.

- The salt/sand mixing/loading area should be swept clean after each load.

- The salt/sand mixing/loading area should be bermed to contain the material until it can be cleaned or the mixing/loading is performed inside the salt storage facility or under a roof. Where paved berms have not been installed, Indiana DOT recommends that A windrow of abrasives [sand] should be placed around all outside stockpiles [salt, salt/sand mix piles]. While minimally effective as a deterrent to stormwater/ meltwater run-off, especially from a sloped surface, on level surfaces such windrows can allow pooling that would otherwise sheet flow around stockpiles causing migration of salt-contaminated stormwater off-site.
- Containers of petroleum and liquid wastes stored outdoors are on a roofed pad enclosed by secondary containment.
  - Herbicide and paint mixing and loading should be done in designated, bermed areas, preferably on a pad.
  - Any spill or residue should be immediately cleaned up.
- Right-of-way trash and construction debris should be taken to a permitted landfill and not allowed to accumulate on site.
- Salt bed washing should be performed in a washbay, not outdoors, and salt bed oiling, paint chipping and painting? if performed at the salt bed rack? should be done with the ground protected by a tarpaulin.

Structural BMPs

Structural BMPs are installed at most DOT facilities to reduce the offsite impacts of contaminated stormwater/ washwater migration, though most of these have been added since construction.

- Standard specifications for new salt storage buildings should include sufficient area for:
  - Sand storage and salt/sand mixing/loading indoors
  - Brine making

  - Storage and bulk tank loading outdoors on a pad protected with secondary containment.
- The exterior pad (to the salt storage building) should be sloped away from the building to its outer limits and the water retained by means of a curb or slope reversal of the pad itself in order that the run-off may be directed into a collection system.
  - The design of the exterior pad (where the mixing/loading operations are performed) should be mandated, not recommended, because the lack of exterior pad curbing has created over half of the salt contaminated stormwater problems observed in at least one DOT survey.
  - The curbing should only be used to allow pooling or to direct stormwater to a collection system. It should not be employed to direct stormwater offsite, as a point source discharge.
- Roofs should be extended on old salt domes to provide a protected area for mixing/loading and for replacement of smaller salt storage facilities.
- Design and specification of structural BMPs should include central office staff and adherence to specification, with input from field personnel to avoid stormwater/washwater collection/discharge problems resulting from poor design.

The following practices are recommended for stewardship of water quality, run-off from maintenance yards and potential discharges from floor drains.

FLOOR DRAIN MANAGEMENT

Caltrans, Maine DOT, and NYSDOT utilize some of the following environmental stewardship practices for floor drain management.

- Direct discharges from floor drains to groundwater through leach fields, septic systems, or dry wells should not be allowed.
- Consider whether a floor drain discharge is truly

necessary. If not, plug the floor drains with a plumber's plug or concrete. If the drain is permanently closed and no discharge can occur, a grit collector, oil/water separator and NPDES/SPDES permit would not be required. A permit would be required for a grit collector and oil/water separator that discharge floor drain waters to surface water. Such permits may involve requirements to change vehicle parking patterns and perform vehicle maintenance only in areas away from the floor drains.

- At facilities that do not discharge shop floor drain effluent to a Publicly Owned Treatment Works (POTW), oil/water separators should be used for washwater. Oil/water separators are effective at removing nonsoluble oil and other petroleum products, but do not remove substance in solution, such as antifreeze and chlorides from road salt. At some facilities, the oil/water separator is connected to a tank, catch basin or holding pond where the washbay effluent collects before being conveyed offsite. A hazardous waste or liquids recycling contractor pumps the contents of the separator or the holding tank, when needed.
- Shop floor drains should be segregated from washbay drains and the flow should terminate at the oil/water separator or, beyond, at a holding tank. Shop floor drains are intended to capture any spills of automotive fluids occurring during vehicle maintenance. No other liquids, including washwater, should be allowed to enter the drain.
- Centralize vehicle repair and maintenance at a subdistrict or district shop where possible, to avoid contamination of facility stormwater discharge.
- Where possible install a grit collector and oil/water separator and connect to a municipal sewer system which eliminates the need for a NPDES/SPDES permit.

- All floor drains should be are constructed with an oil/water separator as part of the system. All floor drain effluent must be forced to pass through an oil/water separator prior to being discharged from the system (before flowing to a tank, sewer district pipe, or infiltrated onto the surrounding grounds).
- Division Managers/Engineers are responsible for knowing where floor drains discharge, and for compliance with the Department Floor Drain Policy. Division Engineers/Managers should keep a current database of the method for managing wastewater from garage floors. This database should include, at a minimum, the presence and type of construction for floor drains and the method of managing effluent. Any changes to the construction or management of this effluent must be reported to the Highway Maintenance Engineer.
- Hazardous waste containers should be provided with secondary containment where risk of damage is high (such as from vehicles) or where impacts from a spill would be severe ' such as spills to floor drains that discharge to the ground. Secondary containment must be capable of holding 110 per cent of the waste container volume.
- Weekly inspections of hazardous waste storage areas should be performed.
- An Emergency Action Plan should be developed in accordance with Bureau of Labour Standards/OSHA (29 CFR 1910.38) requirements for Hazardous Waste Operations and Emergency Response (29 CFR 1910.120(a)(i)(v)) state Emergency Response Planning Procedures for each facility storing hazardous waste.
- Good spill prevention practices should be used. These include, at a minimum:
  - Keeping waste containers closed when not in use;
  - Protecting containers from damage from vehicles or other equipment;

- Use of containers that are in good condition (not severely dented or rusty);
- Use of funnels when pouring liquids into waste containers; and
- Conducting periodic inspections of waste storage area

Management of Oil/Water Separators

Oil/water separators are tanks that collect oily vehicle wash water that flows along corrugated plates to encourage separation of solids and oil droplets. The oily solids or sludge can then be pumped out of the system through a different pipe. The sludge can be hauled off site, and the wash water can be discharged to vegetated areas or to a treatment plant.

There are two types of oil/water separators, one that removes free oil that floats on top of water, and one that removes emulsified oil, a mixture of oil, water, chemicals, and dirt. Choose the separator that fits the needs of the vehicle wash facility. Each oil/water separator should be cleaned of all liquid and grit at least annually.

- Once all free-floating petroleum products are absorbed, the liquid may be decanted to the municipal sewer system or to a tank, for final disposal at a waste water treatment facility or hazardous waste location.
- If there is reason to believe that hazardous materials, other than oils, are in the liquid portion of the oil/water separator, then the Division Engineer/Manager must be notified and arrangements made by a licensed hazardous waste contractor to collect the liquid.
- All grits are to be treated as special waste and disposed of at a special waste landfill (Norridgewock) Hampden, etc.
- Once empty, oil/water separators must be filled with clean water above the bottom of the outflow pipe.
- Oil-only absorbent materials are to be used to capture free-floating oils that may accumulate in the oil/water separator.

Management of High Risk Effluent

Stewardship practice calls for both high risk and low risk effluent to be managed in one of the following ways:

- Discharged directly to a municipal sewer system, with knowledge and permission of the local district.
- Captured in a tank, then transported off-site for final disposal.
  - Highway/Bridge/Traffic Superintendents are to make arrangements for the transport and disposal of the tank contents with either a waste water treatment facility or hazardous waste contractor. Superintendents make arrangements for any analytical testing of the effluent, as may be required by the waste water treatment facility or hazardous waste contractor.
  - Within 4 days of a tank alarm sounding, Crew Supervisors should notify their Superintendent that the contents of a tank need disposal.
  - Bills of Lading, manifests, or other receipts for disposal should be kept at the Division Office for a minimum of 3 years.
- Crew Supervisors should test the tank alarm system monthly. A log of such tests will be kept on-site for a minimum of three years.
- The floor drain can be eliminated and the shape of the floor modified such that no liquids exit the garage through a drain system.

Further options exist and are detailed below for management of low risk effluent.

Management of Low Risk Effluent

In addition to the options available for high risk effluent, low risk effluent can also be managed in one of the following ways:

- When Division Engineers/Managers choose to manage floor drain effluent by separation of activities and bays, the Division Engineer/Managers are to ensure that garage areas where activities could create High Risk liquid effluent are physically separated from the areas where activities create Low Risk liquid effluent. Physical separation includes walls and concrete or polyethylene berms. Berms should also be constructed in a manner that does not allow any waste that could be considered high risk to enter the areas where liquid effluent could be considered low risk. These berms must be constructed in a way that meets OSHA standards for tripping hazards (meet slope requirements of ADA and are striped as a hazard).

- Captured in a melt-water only holding tank, pumped to the surface of the ground, and allowed to infiltrate only if the following conditions are met:
  - Connection to a municipal sewer system is not an option.
  - The infiltration area must be accessible for inspection.
  - The effluent must not discharge directly into a ditch, stream, wetland, pond or other surface water body.
  - There must be no significant potential for pollutants to drip, leak, spill or wash into the floor drains from which the effluent originates. Engine maintenance activities are prohibited from areas which feed floor drains discharging to pipes on top of the ground or melt-water only holding tanks. All containers of oils, engine fluids, cleaning products or other liquid pollutants must be removed or separated by means of an impermeable berm from the area containing the floor drain.
  - Oil-only absorbent materials are to be used to capture free-floating oils that may accumulate in the melt-water only holding tank.
  - Discharges must be done in a way that prevents erosion and the creation of pools of standing

water. Discharge onto frozen ground is not allowed.
  - Effluent must not be discharged within 300 feet of a private well or water intake or within 2500 feet of a public well or water intake.
  - If there is reason to believe that the contents of the melt-water only holding tank has received oil, diesel or other hazardous materials due to a spill or leak, then the contents cannot be discharged on the surface of the ground. The contents of the tank must them be treated as high risk effluent and disposed of in a manner as described in the previous section.
- Discharged directly from the floor drain onto the surrounding ground (daylighted) only if the following are met:
  - Connection to a municipal sewer system is not an option.
  - The pipe must discharge on top of the ground to an infiltration area that is accessible for inspection.
  - The pipe must not discharge directly into a ditch, stream, wetland, pond or other surface water body.
  - There must be no significant potential for pollutants to drip, leak, spill or wash into the floor drains. Engine maintenance activities are prohibited from areas which feed floor drains discharging to pipes on top of the ground or melt-water only holding tanks. All containers of oils, engine fluids, cleaning products or other liquid pollutants must be removed or separated by means of an impermeable berm from the area containing the floor drain.
  - Proper erosion control methods are used at the end of the pipe.
  - Effluent must not be discharged within 300 feet of a private well or water intake or within 2500 feet of a public well or water intake.

Floor Drain Maintenance and Sludge Removal

- Supervisors at garages that contain floor drains are to ensure that floor drains are cleaned a minimum of once per year. The residue removed from the floor drains is assumed to be special waste. Bills of Lading for this disposal are to be kept at the Division Office for a minimum of three years.
- Crew Supervisors should be responsible for ensuring that all floor drains, oil/water separators, and melt-water only holding tanks have oil absorbent socks maintained in them at all times. These petroleum socks, placed in the floor drains, oil/water separators, and melt-water only holding tanks, are to be changed when they show evidence that oil has been absorbed.

Holding Tanks and Maintenance

Environmental stewardship practice for holding tanks may include:

- The minimum holding tank capacity must be 1,000 gallons.
- Holding tanks and piping must be watertight and sealed with materials compatible with the liquid or sludge being stored.
- Access must be provided to each compartment of the tank for inspection and cleaning by means of either a removable cover or manhole (minimum diameter 20 inches). Manholes must extend to finished grade.
- The tank must be designed for the expected maximum structural load and ballast must be provided when necessary to prevent structural damage when the tank is emptied.
- The volume between inlet cover and the maximum water depth must be equal to approximately 20 per cent of the liquid volume stored below the maximum water depth. An alarm with both visual and audio signals must be activated once the water level reaches the maximum water depth.

- A year-end record of pumping events should be produced each year the holding tank is in operation.

Leaks into Floor Drains

If, in the unlikely event, a spill or major leak occurs that allows petroleum or antifreeze liquid to enter a floor drain that was considered low risk, steps must be taken immediately to insure that none of the liquid contaminates the environment through the daylighted pipes, or, tank overflows. If contamination occurs, the steps for major spills must be followed.

VEHICLE WASHING

After a storm, equipment cleaned to reduce corrosion damage and to prepare for the next storm. Water used in washing cars, trucks, and other vehicles may contain a wide range of contaminants, including oil, other hydrocarbons, metals, detergents, road salt, and grit. Discharged into surface waters, these contaminants can degrade water quality and harm aquatic life.

Discharged into groundwater, potable water may be rendered unpotable. Various states are currently conducting studies designed to test the effectiveness of various structural BMPs and investigate the costs of design or purchase, as well as installation and maintenance.

State DOTs are exploring some of the following options and environmental stewardship practices:

- Reduce the sources of potential contamination throughout the state by centralizing saltbed and truck washing at facilities already connected to POTWs amenable to accepting brine discharge. This involves extra driving, and POTWs to which washing facilities are connected may have imposed limits on the volume of washwater or the levels of chloride and cyanide discharged from the facility. Washing facilities may be at such a distance from the others that moving trucks for washing is inconvenient or impractical, regardless of the cost; however, when usable facilities are accessibly located, this is a cost effective option.

- Connect every truck washing facility to an amenable POTW. Costs of extending the line can be shared with other dischargers in some cases, though some facilities are too remote to make this a practical option.
- Contain wash water in holding tanks and haul to an amenable POTW. INDOT found this is the most cost effective option for facilities that
  - Cannot afford to connect to a POTW;
  - Cannot discharge washbay effluent to a POTW because of prohibitions;
  - Will install brinemaking equipment and can use washwater as "make-up" solution or
  - Will continue to spread road salt for all or some snow/ice events.

  Vacuum trucks or bulk tank trucks (with a pump), if available, can be used to reduce hauling costs. While some POTWs will not accept transported liquids, a pick-up service might be arranged where amenable POTWs can be located.
- Line existing and newly constructed holding ponds with a clay layer or plastic impervious liner and design the structure to hold the maximum volume of meltwater, washwater and precipitation that can conceivably collect while evaporation is depended upon to reduce the volume (no overflow is allowed).
- Install catch basins, settling tanks and holding ponds to remove suspended particles. Dissolved chlorides cannot be removed in this fashion. According to an INDOT/Purdue study, the holding pond should be sized to hold the maximum volume of washbay effluent, surface area stormwater/meltwater run-off (unless the pond is bermed) and precipitation (roofing over a holding pond retards evaporation). Difficulties are encountered removing sediment from any holding pond, but especially one lined with plastic. Unless

properly maintained, holding ponds can collect debris and serve as a harbor for algae blooms, wild fowl and reptiles. The cost of constructing a holding pond as a BMP must be weighed against the costs of cleaning and maintenance and the potential for groundwater contamination via a perforated or breached liner. Also, evaporation cannot be relied on to reduce the total volume contained because it is periodically replenished by precipitation. Prohibiting any discharge from a pond will mean that the contents will need to be pumped and hauled to an amenable POTW every other year; a longer cycle will increase the probability that no POTW would accept the contents because of the increased brine concentration. Alternatively, the unevaporated content can be pumped to tanks, if available, and used as brine makeup solution or hauled to another facility for this purpose.

Design and Operation of Washing Facilities

New Hampshire, Oregon, and other states have identified environmental stewardship practices for vehicle washing facilities.

The following environmental stewardship practices can be added to existing and or new DOT facilities to minimize the potential for environmental contamination from polluted run-off:

- Warning signs should be posted for customers and employees instructing them not to dump vehicle fluids, pesticides, solvents, fertilizers, organic chemicals, or toxic chemicals into catch basins. Catch basins are chambers or sumps which collect run-off and channel it to the stormwater drain or to the sanitary sewer. Vehicle wash facilities should stencilwarnings on the pavement next to the grit trap or catch basin. All signs should be in a visible location and maintained for readability.
- Care should be taken to minimize wash water run-off from cleaning operations.

  - Minimize water use to reduce potential for unpermitted non-stormwater discharges.
  - When possible, truck beds should be cleaned using a dry cleanup technique (sweep up or shovel out).
  - Use low pressure (brush and hose with nozzle only. no power booster or steam cleaning)
  - Exterior and frame wash only
- No detergents should be used, as they emulsify oil in the oil/water separator and make the separator ineffective and may violate a DOT's NPDES/SPDES permit by introducing new chemicals. Using alternative cleaning agents such as phosphate-free, biodegradable detergents for vehicle washing can also reduce the amount of contaminants entering storm drains.
- Where possible, indoor wash facilities with controlled floor drainage should be utilized. Where wastewater is not to be disposed to a sanitary sewer, grassed swales (shallow) vegetated ditches, or constructed wetlands (retention ponds with emergent aquatic vegetation) can be used to hold wastewater and allow contaminant removal through infiltration and filtration.
- Washwater may otherwise be collected in a sump, grit trap, or containment structure to be pumped or siphoned to a vegetated area so that complete percolation into the ground occurs. A portable collection system will provide the collection of the contaminants provided the collection system is large enough to capture significant amounts of the overspray. Washwater run-off can also be disposed of into an infiltration basin/trench. The use of a bioswale with an oil/water separator will virtually eliminate the total suspended solids, oil and grease, and heavy metals discharged provided both are properly sized.
  - Disposal of washwater should occur on ground surfaces with vegetated cover, preferably grasses,

a minimum of 250 feet in length before a surface water body. A distance of 250 feet was based on a hydraulic conductivity of 0.2 gal/ft/day, volume per day of 150 gallons, and a swale with a width of 3 feet.

- Complete percolation in the swale should occur with no direct discharge to the surface water. Discharge into a grassy swale for treatment should not occur within 24 hours after a rainfall event or if water remains ponded in the swale.

- Wash areas should be located on well-constructed and maintained, impervious surfaces (concrete or plastic) with drains piped to the sanitary sewer or other disposal devices. The wash area should extend for at least four feet on all sides of the vehicle to trap all overspray. Enclosing wash areas with walls and properly grading wash areas prevent dirty overspray from leaving the wash area, allowing the overspray to be collected from the impermeable surface.
- The impervious surfaces should be marked to indicate the boundaries of the washing area and the area draining to the designated collection point.
- Washing areas should not be located near uncovered vehicle repair areas or chemical storage facilities; chemicals could be transported in wash water run-off.
- Regularly inspect and maintain the designated areas, facility wash racks, designated cleaning areas, wash pads, clarifiers, oil-water separators, sumps and sediment traps. Regularly clean wash areas, grit traps, or catch basins to minimize or prevent debris discharge, such as paint chips, dirt, cleaning agents, chemicals, and oil and grease into storm drains or injection wells.
- A washwater treatment sequence may include such elements as a grit trap, an oil/water separator, a dosing tank with siphons or pumps, and a multi-media filter bed with underdrains.

  - Discharge from underdrains must meet effluent limitations.
  - Maintenance of a multi-media filter should consist of cleaning, removing the top inch of sand once every six months; when the total depth of filter sand fall below 18-inches, the sand should be replaced. If clogging and/or short circuiting occurs as observed by uneven infiltration in the filter or formation of surface cracks, the sand should be replaced.
  - A Spill Prevention, Control and Countermeasure (SPCC) Plan, in accordance with 40 CFR 112, should be prepared and implemented to prevent the entry of pollutant loads beyond the capabilities of the treatment system.
- Contractual provisions should require contractors to use cleaning practices consistent with DOT requirements.

Recycling Wash Water

Recycling systems reduce or eliminate contaminated discharges to stormwater drains and injection wells by reusing the wash water until the water reaches a certain contaminant level. The wastewater is then discharged to a collection sump or to a treatment facility. Collection sumps are deep pits or reservoirs that hold liquid waste. Vehicle wash water accumulates in the collection sumps, and is pumped or siphoned to a vegetated area (such as a grassed swale or constructed wetland). Sediment traps can also be used to strain and collect the vehicle wash water, prior to pumping or siphoning the wash water to a vegetated area.

The use of a recycling system can reduce or eliminate the contaminant discharge to stormwater or sanitary sewer while greatly reducing the amount of water used in the process. Some DOTs are installing brinemaking at truck washing facilities so washwater can be used as "make-up" solution. This solution can also be used to "spray the load" of salt/sand mix and/or to fill saddle tanks for spraying the mix as it passes through the salt spreader. The cost of brinemaking equipment is relatively affordable, though adding brine application equipment to existing trucks is more costly.

A recycled wash water has been successfully used by one of the largest bus transport companies operating in Borsod County, Hungary. In 1985, they installed a new, water-saving wastewater treatment facility for wastewaters resulting from washing at the central service plant. The commercial transportation system uses detergent-free, high pressure, hot water to remove dirt and grime from the car bodies and engines of the buses. The resulting wastewater is mechanically treated with filters consisting of sand and activated carbon. For disinfection, a 1 to 3 mg/l NaOCl solution is used.

The filters are backwashed with recycled water every 3 to 4 days. The polluted backwash water is returned to the treatment plant. Oily rainwater from the yard is also directed into the treatment plant.

The system uses fine sand filtration after pretreatment of the wastewater to remove grit, sand and oil. After this pretreatment, about 15 per cent to 20 per cent of the wastewater is discharged into a conventional sewerage system. This discharge prevents the accumulation of TDS and organic substances in the remaining water which is recycled for use in the carwash. This discharged water meets the water quality requirements for all categories.

The remaining water that is to be recycled is subjected to ozonation to prevent anaerobic digestion of organic materials which produces foul odours. After ozonation, the remaining, pretreated water is conveyed through a fine sand filter by pump. Once filtered, the water is resupplied to the carwash by means of a rubber membrane hydrophore at a pressure of between 2 and 8 bar. For the commercial vehicle washing recycling facility, the initial investment costs are about $80,000, with a further investment of $1,600 for reconstruction after about ten years of operation. Maintenance costs were about $4,000/year. The estimated period for recovery of this investment is about 1.3

years based upon typical usage within the region. The technology achieved 80 per cent recycling of water.

In addition to the recycling of washwater, environmental stewardship practices associated with recycling washwater include:

- Recycling treatment equipment should be properly operated and maintained to achieve compliance with all conditions of the permit.
- Backwash water or concentrate water should be properly discharged to sanitary sewer.
- Liquid concentrate discharged to the sanitary sewer should meet all pretreatment standards and other requirements of the local Sewer Authority.
- Solids, grit, or sludge should be disposed in a manner that complies with State administrative rules.

ENERGY CONSERVATION

Nova Scotia, Canada is one of the North American transportation agencies on the state or province level that has taken a serious look at energy conservation practices on the facility level. The Nova Scotia Department of Transportation and Public Works (NSTPW) reports in their practice guide that in 1996, electricity generation accounted for 40 per cent of Nova Scotia's greenhouse gas emissions. Nova Scotia is participating with the federal government and the other provinces and territories in developing the National Climate Change Strategy for Canada.

Towards this effort and implementation of the agency's EMS, NSTPW is seeking to:

- Reduce activities which use electricity
- Improve efficiency both in the distribution of electricity and in the consumption of electricity by energy users
- Use less carbon-intensive forms of electricity generation.

Energy use at TPW locations can be broken into four categories: HVAC systems, lighting, water heating, and office equipment. Space and water heating alone was estimated to be responsible for up to 35 per cent of energy use in office buildings.

The New York City Transit Agency has been a national leader among transportation and other public agencies in the U.S. in Green Building design and operation, which includes design and operation for energy efficiency.

PLANNING FOR ENERGY CONSERVATION

- Conduct an energy audit to determine how much energy is being consumed at your location; break down the results by category of energy use if possible.
- Establish goals for the overall energy consumption of the building or renovated area. These goals should be broken down by category of energy use. Inform employees of these goals.
- Organize electrical services to permit sub-metering of energy use by category: cooling, pumping, fans and heating, plug loads, etc.
- Collect and analyse sub-metered energy data on a regular basis; compare results to goals for each category.
- Consider alternative heating sources such as ground-source heat pumps, geothermal heat pumps, solar or other renewable energy sources.
- Provide employees/occupants with information on how they use energy and what they can do to reduce their energy use.

HEATING, VENTILATION AND AIR CONDITIONING 'HVAC,

- Address leaks and poor insulation in HVAC ductwork.
- Develop operating manuals for all equipment including design intent, set points, setback and setup schedules, on/off time schedules, special features and requirements, etc.
- Develop maintenance manuals for all HVAC equipment with schedules and frequency of service required.
- Ensure heating and cooling are on different schedules

and that it is not possible to have the two operating coincidentally.

- As office schedules change, ensure time schedules for ventilation fans, purge cycles, heating and cooling are changed to match building occupancy.
- Provide preventative maintenance checks annually to ensure all HVAC systems are operating properly; make any necessary repairs.
- Carry out an annual calibration and check the function of all building automation systems to verify operation and performance.
- When demand controlled ventilation systems are used, carbon dioxide sensors must be calibrated annually to ensure proper function.
- Ensure service technicians provide detailed listings of all service performed and findings made. Ensure all changes to equipment are documented and all parties affected are informed.
- Records for inspections and repairs should contain:
  - Date of inspection and/or repair.
  - Inspection company and/or persons contact information.
  - Details of work completed, including costs.
  - Date for next inspection.
  - Any malfunctions of system found during inspection and/or repair.

LIGHTING EFFICIENCY

Technologies developed during the past 10 years can help cut lighting costs 30 per cent to 60 per cent while enhancing lighting quality and reducing environmental impacts.

- Turn lights off when not needed and ensure that occupancy sensors have not been overridden.
- Have the cleaning and maintenance schedule overlap with regular hours to minimize energy use. After hours work should be done by area, using only necessary lights 'task lighting.

- Introduce local task lights, allowing a reduction in general overhead lighting.
- Keep light fixtures clean as dust greatly decreases the amount of light delivered.
- Specify single bulb and fluorescent, mercury-free fixtures to replace incandescent ones at the end of their useful life; use the most energy efficient lamps and ballasts available for replacements.
- Consider group relamping. Common lamps, especially incandescent and fluorescent lamps, lose 20 per cent to 30 per cent of their light output over their service life.
- Replace all the lamps in a lighting system at once. This will save labour, keep illumination high, and avoid stressing any ballasts with dying lamps. It is useful to keep a record of the type of bulbs used and when they are replaced. This allows for long-term monitoring of the efficiency and life span of different types of lighting in different areas.
- Ballasts-standard choke ballasts can be replaced by high frequency electronic ballasts. Electronic ballasts are highly recommended for use with low-voltage tungsten-halogen lamps, high-efficacy argon-krypton filled fluorescent tubes, metal-halide and high pressure sodium lamps. Electronic ballasts offer the following advantages:
  - 20 to 30% energy reduction compared with conventional ballast
  - 50% longer service life of lamps
  - Absence of flicker 'ballast operates lamp at a frequency between 22 and 70 kHz,
  - Silent operation
  - Net power factor of 0.95 to 0.99
  - Low harmonic distortion
  - Overvoltage protection
  - Automatic switch-off of faulty or end-of-life lamps
  - Reduction in weight
  - Cool operation.

- *Automatic control systems*:
  - Timer circuits that switch lamps off during room vacancy times
  - Photoelectric sensors that sense the amount of daylight in the room and either switch lamps on or off or adjust the lamp brightness accordingly
  - Occupancy sensors that switch lamps off when work stations are unoccupied.

COMPUTER EQUIPMENT

Computers, printers, photocopiers and fax machines are the greatest contributors to office paper waste and energy consumption.

- Enable energy saving options on computers, monitors, printers, photocopiers, etc.
- Turn all equipment off when not in use (both day and night); this will also increase equipment lifetime.
- Batch copy jobs rather than doing single copies.
- Buy printers and photocopiers that can do double sided printing or reduce size to fit.
- Ensure that printer and toner cartridges can be returned and recycled by the manufacturers.
- Choose equipment based on its efficiency and operating costs over time; only buy office equipment that has the Energy Star or Environmental Choice, EcoLabel. Ensure that the Energy Star programme is initiated when equipment is first installed.

UNDER AND ABOVE-GROUND STORAGE TANKS

Above ground storage tanks are used at Maintenance facilities to store fuel, oil, antifreeze, deicing agents and asphalt emulsion.

Underground tanks most often contained fuel. These materials can pose potential threats to water quality if spilled or mixed with stormwater run-off. Many DOTs have removed underground storage tanks for petroleum products, except at major facilities, and replaced these with above ground tanks or credit cards to purchase fuel private sources. Environmental stewardship practices for preventing and addressing storage tank leaks and spills include those reviewed below.

TANK REGISTRATION PRACTICES

Fuel/petroleum tanks are generally registered with the state environmental agency when they can hold in the range of more than 1,100 gallons in a total combined capacity of any aboveground and/or underground tanks which store petroleum products, gasoline, diesel, kerosene, used oil or heating fuel. All fuel/petroleum tanks storing used oil are registered regardless of tank size or site capacity. One exception to this registration requirement—small tanks that are not manifold and hold less than 1,100 gallons of heating oil for on-site use are exempt from this regulation.

- Petroleum Bulk Storage (PBS) Registration certificate should be posted near the tank location and renewed every five (5) years.
- The PBS certificate should be updated whenever a tank removed or modified or added to reflect all current tanks.
- The state environmental agency is notified prior to any tank removals.

TANK EQUIPMENT AND RECORDKEEPING PRACTICES

- In both aboveground and underground petroleum storage tanks colour-coded fill ports are used per the American Petroleum Institute's colour and symbol code
- In both aboveground and underground petroleum storage tanks the colour and symbol code is identified on each fill port
- Typical DOT petroleum storage tank port colours and codes are: Diesel - Yellow; Unleaded Gasoline - White with Black Cross; Kerosene - Brown
- In both aboveground and underground petroleum

storage tanks operating valves are present on all gravity-drained tanks

- In both aboveground and underground petroleum storage tanks automatic shut-off valves such as solenoid valves on gravity-fed systems and shear valves on pumped dispensing systems are present to prevent any leaks in case of a pipe or hose failure
- In both aboveground and underground petroleum storage tanks check valves are present for backflow prevention on tanks filled by pumping
- For underground petroleum tanks monitor the inventory by measuring use, deliveries, losses or gains, and bottom water levels daily.
- For underground petroleum tanks record inventories to the nearest tenth of a gallon if feasible,
- For underground petroleum tanks reconcile records every ten days and retain all records for at least five years.
- For underground petroleum tanks report inventory losses or gains of product that are more than:
  - 0.75 per cent of the tank volume in a ten day period
  - 7.5 gallons for every 1,000 gallons delivered over a ten day period or 0.75 per cent of the throughput or amount dispensed over a ten day period
  - Or if water seems to be accumulating during any ten (10) day period
- When the cause of losses or gains in underground petroleum tanks cannot be explained within 48 hours, the Department of Environmental Conservation's Regional Spill Engineer is contacted and the tank is taken out of service
- Tightness testing is required every five (5) years when underground tank systems are not protected and hold more than 1,100 gallons. (Protected systems are those that were installed with corrosion-resistant tank and piping and a leak monitoring system)

- For underground petroleum tanks monitor corrosion-resistant tanks and pipes at the manufacturer's recommended schedule.
- Steel USTs with corrosion protection should comply with the following requirements to ensure that releases due to corrosion are prevented for as long as the UST is used to store regulated substances.
- For underground petroleum tanks inspect leak detection systems or double-walled tanks at least weekly, and other monitoring systems monthly.
- For underground petroleum tanks check any cathodic protection systems at least annually.
- Tanks (tanks for on-site heating are exempt) that hold 110 gallons or more are required upgrade to meet federal underground tank storage requirements. These federal standards included leak detection systems, corrosion protection, and spill/overfill prevention features.
- For aboveground petroleum tanks inspect monthly the exterior surfaces of tanks, pipes, valves, leak detection systems and other equipment to identify any cracks, wear, corrosion, settling, separation or other problems. Keep records with the date and signature of the inspector for ten (10) years.
- For aboveground petroleum tanks perform secondary containment and internal inspections every ten '10, years conducted by qualified firms for any tanks of greater than 10,000 gallon capacity that rest on the ground. Secondary containment may be required for smaller tanks that could reasonably be expected to discharge to waters of the state.
- For aboveground petroleum tanks gauges accurately showing the product level are present unless a high level alarm or a cutoff controller is present. The design and working capacities and tank identification number must be clearly marked on the tank and at the gauge.
- Erosion is the main concern with earthen emergency

spill containment. Dried weeds and grass present a fire hazard. Animals can burrow through earth dikes. Concrete structures are susceptible to cracking and frost damage. A weekly inspection schedule should be developed to address these problems so they can be repaired promptly. Individual site-specific checklists should be developed to reflect site specific concerns. The following is a list of key items to be addressed in the periodic inspection of ASTs. Each AST should be inspected at least weekly.

- Presence and/or volume of oil or water in the containment area
- Soil colour changes; noticeable sheen on water puddles
- Visual observance of tanks, pumps, valves, and pipe connections
- Unusually strong odour of stored material
- Storage tank overflowing
- Determination of accumulated liquids contained in area (uncontaminated stormwater, contaminated run-off, or pure product).

• New UST systems should be properly installed in accordance with industry codes of practices. At a minimum, new underground storage tank systems including piping should be:
  - Properly designed, constructed, and protected
  - From corrosion
  - Equipped with leak detection devices
  - Equipped with spill/overfill protection
  - Installed by a state certified installer
  - Inspected by a state certified inspector during installation
  - Registered with state upon installation.

• Record keeping is necessary to ensure compliance with the federal, state, and local regulations. Records of a tank system should be maintained for the operating life of the system and at least 5 years after its permanent closure. The original documents should be maintained on site, if possible. The following records are often necessary to document appropriate environmental stewardship:

- Inventory Records
- Installation Details
- Modification/Repair Details
- Operation Records
- History of usage
- Physical inspection checklists/reports
- Monitoring records
- Leak/incident documentation
- Tank Handling Activity Report
- Legal Records
- Permits, notifications, and certificates
- Agency correspondence
- Consultant/Contractor Reports
- Engineering assessments/surveys
- Tank and line testing results
- Environmental sampling results
- UST Inspection Report
- Tank Closure Records
- UST Closure Notification Form
- UST Closure Report Form
- Registration of Storage Tank Form
- Notification of Reportable Release/Notification of Contamination

- Prior to releasing rainwater from secondary/spill containment of an AST, inspect the water for contamination. If there is evidence of spilled or leaked material, or captured rainwater in the spill containment exhibits a surface sheen, contact the District Stormwater Coordinator or District Hazmat Manager for appropriate actions to be taken.
- Make sure the drain valve or plug is properly secured to contain any future leaks or spills.

- Be prepared to respond in the event of a leak or spill from an above ground tank:
  - Maintain an appropriate spill kit near each above ground tank.
  - Replenish spill kit supplies as they are used.
  - If the type of material being stored changes, replace spill kit contents with supplies appropriate for the new material.
- For cleaning spilled materials, particularly hazardous materials, follow procedures from the product MSDS (Material Safety Data Sheet) or the *North American Emergency Response Guidebook* (Federal DOT document).
- For facility-specific procedures related to spills, refer to the *Spill Prevention Control and Countermeasures* (SPCC) Plan, Title 40, Code of Federal Regulation, part 112 or the Hazardous Materials Business Plan.

VEHICLE AND EQUIPMENT FUELING PROCEDURES AND PRACTICES

Vehicle and equipment fueling procedures and practices are designed to minimize or eliminate the discharge of fuel spills and leaks into stormwater drainage systems or watercourses during equipment fueling and the bulk delivery of fuel.

BULK FUEL DELIVERY

- All aboveground and underground storage tanks should be equipped with automatic overfill shutoff valves.
- Spill Prevention and Control BMPs should be implemented to prevent spillage.

FUELING AREA MAINTENANCE

- Label drains at fuel dispensing areas to indicate if they discharge to the storm drain or to the sewer.
- Storm drain inlets may be temporarily covered with spill pads and/or mats during fueling operations.
- Absorbent spill cleanup materials or drip pans should be stored in fueling and maintenance areas and used

materials should be disposed in accordance with hazardous waste management BMPs.

- Immediately clean up leaks and drips.
- Hosing off the fueling area is prohibited. Dry shop clean up practices should be used.
- Manage wastes to reduce adverse impacts on stormwater quality. Fueling areas should be kept free of litter and debris that might become contaminated with petroleum products.
- Maintain and implement a current spill response plan for fueling operations.
- Inspect fueling facilities daily and correct deficiencies.
- Keep a supply of spill cleanup materials on site.

REFUELING PRACTICES

- Fueling operations should not be left unattended. Fueling in the field should not be performed near unprotected drainage facilities or watercourses.
- Drip pans should be used during vehicle and equipment fueling unless the fueling is performed over an impermeable surface in a dedicated fueling area. Dedicated fueling areas should be protected from stormwater run-on and run-off and should be located at least 15m from downstream drainage facilities or watercourses.
- Nozzles used in vehicle and equipment fueling should be equipped with an automatic shutoff to control drips.
- Warnings against "topping off" fuel tanks should be posted at fuel dispensers.
- Fueling operations should not be left unattended.
- Absorbent spill cleanup materials should be available in fueling and maintenance areas and should be disposed properly after use.
- Vehicles and equipment leaks should be inspected and cleaned up on each day of use.

- Leaks should be repaired immediately or problem vehicles or equipment should be removed from the project site.

A SPCC Plan outlining procedures and measures to prevent and respond to a petroleum spill is required if:

- The underground tanks at the facility can store more than 42,000 gallons
- 660 gallons or more can be stored in a single aboveground tank
- 1,320 gallons or more can be stored in some combination of smaller aboveground tanks,
- Portable containers and oil filled equipment or transformers.

The potential for groundwater contamination raises the need for action at a site to a high priority level. The RCRA environmental indicator for controlling migration of contaminated groundwater requires the following documentation:

- Consideration of all available, relevant or significant information on known and suspected releases to the groundwater at the facility;
- Determination whether groundwater is contaminated above appropriately protective levels 'applicable promulgated standards, other appropriate standards, guidelines, guidance, or criteria, anywhere at, or from, the facility;
- Determination whether the migration of contaminated groundwater has stabilized (remains within the previously determined existing area of contamination);

Environmental stewardship of suspected groundwater contamination from run-off at maintenance sites or other DOT facilities should include the following, which parallel environmental indicators used by EPA to measure progress towards groundwater remediation at RCRA sites subject to corrective action.

- Where groundwater contamination is known or suspected, the DOT controls the migration of contaminated groundwater plumes, through:

- Consideration of all available, relevant or significant information on known and suspected releases to the groundwater at the facility.
- Determination whether groundwater is contaminated above appropriately protective levels applicable promulgated standards, other appropriate standards, guidelines, guidance, or criteria, anywhere at, or from, the facility.
- Determination whether the migration of contaminated groundwater has stabilized (remains within the previously determined existing area of contamination).
- Determination whether contaminated groundwater discharges to surface water.
- Determination whether any discharge of contaminated groundwater to surface water is "significant" (the maximum concentration of the contaminant in the surface water is more than ten times the appropriate groundwater level).
- Determination whether the discharge of contaminated groundwater into surface water is "acceptable" until a full assessment and a final remedy decision can be made. Factors to be considered in the interim assessment include surface water body size, flow, use/classification/ habitats, contaminant loading limits, other sources of surface water/sediment contamination, effects on ecological receptors (via bio-assays, benthic surveys or site specific ecological risk assessments performed by trained specialists).
- Decision whether groundwater monitoring measurement data and surface water/sediment/ ecological data will be collected in the future to verify that contaminated groundwater has remained within the existing area of contaminated groundwater.

- Where groundwater contamination is known or suspected, the DOT controls human exposure to contaminated groundwater, through:
  - Consideration of all available relevant or significant information on known and suspected releases to soil, groundwater and surface water at the facility.
  - Determination whether the soil, groundwater or surface water is contaminated above appropriately protective risk-based levels.
  - Determination whether there are complete pathways between contamination and human receptors such that exposures can be reasonably expected under the current land, groundwater and surface water use conditions.
  - Determination whether the exposures resulting from the complete pathways (above) are "significant." Here, "significant" means: (1) greater in magnitude (intensity) frequency and/or duration, than assumed in the derivation of the "acceptable levels" used to identify the contamination, or (2) the combination of exposure magnitude (perhaps though low, and contaminant concentrations) which may be substantially above the "acceptable levels" that could result in greater than acceptable risk.
- The design and implementation of Best Management Practices is a DOT priority at sites where contaminated groundwater is suspected, in order to prevent further degradation and, possibly, migration to surface water, including sources of drinking water.

This documentation meets or exceeds that required by the Phase I Environmental Site Assessment Standard (ASTM E 1527-00), which may be utilized at some sites, and can serve as the foundation for further, Phase II investigations of "recognized environmental conditions" (RECs) identified during Phase I assessments.

SERVICE INSPECTION, MAINTENANCE, AND REPAIR

INTRODUCTION

A key feature of GRS–IBS is that it has fewer parts than conventional bridges and abutments and should therefore need less maintenance. Like other bridges, the main components are the superstructure and the substructure. The superstructure is the same as a conventional bridge and should have the same protocol for inspection, rating, maintenance, and repair. IBS is also somewhat similar to an integral abutment in how the ends are embedded in the approach. A difference is that the IBS is embedded in compacted gravel, whereas an integral abutment is encased in concrete. Both bridges are designed without a joint to limit the effect of water at the beam ends for improved durability. At the time of this report, nearly 30 GRS–IBS bridges had been built for local or off–system service, with the oldest built in 2005. None of the bridges show any signs of distress. All indications are that GRS–IBS works well in the local road environment, suggesting that the long–term performance of this system is adequate. In addition, IBS has fewer components and is designed for a smooth transition, thereby reducing impact loads (a major contributor to fatigue of the superstructure). This promise of improved performance, however, does not mean that the bridge system is immune to the common problems of conventional systems. Focuses on potential requirements unique to this IBS and other components associated with the integrated approach.

IN–SERVICE INSPECTION

Both superstructure and substructure elements should be included as part of the visual inspection process. As previously indicated, the superstructure is similar to a conventional bridge and therefore has a similar procedure for inspection.

The following elements should be included as part of the inspection of the IBS substructure:

- *Pavement*: If the bridge has asphalt pavement, check

for a transverse crack, shoving, or separation at the approach end wall interface.

- *Approach*: Check the approaches for vehicle rideability and smoothness.
- *Parapet walls*: Check the interface between the beams and the parapet wall for separation or shifting.
- *Beam ends*: Check embedded beam ends for corrosion rust stains, at the beam bases.
- *Scour*: Monitor GRS walls built adjacent to a water channel for scour. Riprap or other appropriate countermeasures should be monitored at each bridge inspection or after an extreme flood event. Any movement of rock should be noted and repaired to prevent scour from progressing and endangering the RSF or the abutment. No problems have been noted in installations of GRS abutments even after sequential flooding events. An indicator of scour on an abutment face or wing wall can be achieved by using coloured blocks on the bottom five to eight rows. Solid blocks are recommended at the bottom, as they are more likely to resist any impact of moving riprap, ice, or other abrasion associated with the normal water elevation. The coloured solid blocks are also covered from view by the initial riprap, and any exposure of coloured block during inspection serves as a visual check for movement or undermining of the riprap, indicating a need for remediation or repair to protect the RSF and abutment from scour.
- *Drainage*: All GRS structures should include consideration for surface drainage. Check the critical drainage paths where the fill slope meets the wing walls leading to the base of the wall. It is imperative that the wing walls have sufficient embedment to prevent erosion due to roadway run-off.
- *Wall face cap*: Inspect the coping for cracks.
- *Modular blocks*: For GRS walls built with modular facing blocks, check for the following:

- Cracked blocks.
- Separated blocks.
- Block durability problems. Refer to Chan *et al.* for about block durability due to freeze thaw, spalling, and efflorescence.

• *Guardrail*: Inspect traffic barriers for damage.

• *Wall face*: Inspect wall faces for excessive lateral movement or settlement.
  - Lateral deformation can be checked visually or with a plumb bob referenced from known points from the top of the wall to the bottom.
  - Visual inspection for wall settlement can be achieved by checking for distortion between the horizontal courses of block.

• *Clear space*: Inspect, measure, and record the distance from the top of the wall face to the base of the superstructure beam for any settlement within the GRS abutment mass. A clear space must be maintained throughout the life of the bridge to prevent loading on the facing elements.

• *Drip edge*: Inspect any drip edge detail at the top of the wall beneath the beam for water diversion.

• *Burrows*: Inspect and remove any animal burrows adjacent to the walls.

MAINTENANCE

If properly designed and constructed, GRS–IBS should need minimal maintenance because it has fewer parts (approach slab) sleeper slab, CIP parapet walls, bridge bearings, or joint details. Since the bridge superstructure is built with common materials, general maintenance should be similar to that of a conventional bridge system.

Maintenance duties might include the following:

• Sealing of a pavement crack, particularly one forming at the beam approach interface.

• Stabilization of drainage ditches to prevent erosion along the wing wall.

- Removal of vegetation growth from the wall face unless it is part of the design.
- Sealing of any gaps in the facing large enough to allow for a loss of fill.

REPAIR

This section includes tips and suggested methods of repair in the event of damage to the GRS abutment wall face. Damage can occur as result of impact, unforeseen scour, or poor wall face durability. Since a GRS abutment is internally supported, the face is not considered a structural element. However, its integrity is important to ensuring long–term performance of the GRS abutment.

The following are repair procedures for potential problems:

- *Damage to a few hollow–core blocks within the face of the wall*: Chip out the face of the damaged block and replace it with the face of another block. The face piece should be cut slightly smaller and be secured with mortar.
- *Repair of deteriorating facing blocks or scour damage*: Although there is no case history for this, shotcrete can be used to repair the face of a modular block wall. shows a GRS wall built with CMU blocks being used to repair a failed MSE wall. Figure 3.1 shows the same CMU covered with shotcrete. Note that drains were installed at the base of the wall to facilitate the flow of water from the GRS abutment. In some situations, it might be necessary to install vertical strip drains in the face of the GRS wall before applying the shotcrete.

Fig. 3.1 Use of GRS Wall to Repair Damaged MSE Wall.

Fig. 3.2 CMU GRS Wall with a Shotcrete Face.

- *Damage to the top rows of CMU block*: Figure 3.3 show a GRS wall before and after the repair of a rock–fall impact. The repaired section is set slightly back from the original wall alignment. To repair this wall, the boulder was removed, and each soil layer within the damaged zone was excavated. To access the fill, the fabric layers were cut perpendicular to the face and peeled back enough to access all the reinforcement layers within the damaged zone. This process was repeated until the damaged zone was exposed. The exposed zone was rebuilt using the 1–2–3 method explained. one layer at a time, from the bottom up. In areas where the reinforcement was excessively damaged, new reinforcement was spliced in to reestablish the frictional connection. The top courses were then pinned and grouted.

Fig. 3.3 GRS Wall Damaged by Large Sandstone Boulder.

Fig. 3.4 Repair of a GRS Wall after Damage Caused by Rock Fall.

- *Excessive settlement of the beam seat*: While this has not been observed, it is possible that the superstructure could experience excessive movement either due to compression of the GRS abutment or external instability. If the clear space is lost and the superstructure is causing distress to the wall, it is possible to saw a new gap to relieve the pressure. An alternative method would be to pressure grout and elevate the superstructure back to its original grade, which may also require repair to the approach pavement.

4

Quality Control and Quality Assurance

INTRODUCTION

Quality is everyone's responsibility. The quality of GRS–IBS begins with an understanding of the concept of wall strength and the interrelation of components of a GRS system and with sound construction practices. QC consists of implementation, measurement, and enforcement of sound construction practices and field inspection procedures to ensure construction quality as outlined in this manual.

QC also involves the selection of quality materials. The successful completion of a project is dependent on a proper monitoring programme with necessary adjustments at each stage of construction. QC is the responsibility of the builder.

QA is necessary to ensure the finished product meets specifications through inspection, testing, and final acceptance. The process involves constant evaluation of the project activities related to planning, design, development of plans and specifications, construction, and all interactions associated with these fundamental activities. QA can either be the responsibility of the owner agency or a third–party agency.

ROLE OF THE CONTRACTOR

Since GRS is a nonproprietary generic wall system, the

contractor building the wall can be responsible for developing and maintaining a QA/QC plan for project quality. Prequalification based on the procedures outlined in this manual should be a necessary requirement for this type of construction.

TESTING

QC testing performed during construction mainly applies to onsite field testing of backfill material and associated laboratory tests.

Laboratory Testing

Gradation and moisture–density tests will be required for field monitoring of the backfill material. The classification tests and moisture–density tests should follow AASHTO standards for aggregate sampling and testing.

Large–scale direct shear tests or triaxial tests are the most effective methods for determining the friction angle for coarse–grained backfill aggregates. These methods of testing are preferred over the standard direct shear test (AASHTO T236) or smaller diameter triaxial tests that are performed on the minus No. 10 material.

Field Testing

Fill placement and compaction is the predominant construction activity that needs to be monitored in a GRS–IBS project. Field density tests should be performed on each layer. The field test method should be applicable to the aggregate type that is used for the backfill material.

Dense–graded backfill, which consists of State transportation department crushed base course, can be tested with a nuclear gauge. State transportation department density testing procedures can be used. Moisture content should be monitored and controlled prior to fill placement for an effective compaction process.

A procedural (method–based) specification is preferable for the compaction of open–graded fill material, which exhibits a high percentage of void space. Open–graded gravels are not conducive to in–place nuclear density testing procedures as the direct transmission nuclear gauge procedure is difficult to perform (the transmission hole will typically not stay open), and nuclear backscatter testing is not effective due to poor soil/gauge contact. In lieu of density testing for open–graded gravels, maximum density can be achieved with a recommended procedural specification. The procedure can specify three to five passes with a walk–behind vibratory plate compactor near the wall face. Larger ride–on vibratory rollers with greater frequency and efficiency can be used in the core of the GRS mass with fewer passes 3 ft from the wall face.

CONSTRUCTION INSPECTION

Thorough inspection before and during construction will ensure the GRS structure is built in accordance to the plans and guidelines. Inspection requires an understanding of GRS design and methodology. Familiarity and understanding of the drawings is necessary. It is important to have firsthand knowledge of the GRS construction processes. A properly implemented field inspection programme provides an opportunity to take corrective action during the construction process.

A critical component of construction is compaction behind the facing element followed by placement of the geosynthetic reinforcement. Those responsible for performing these construction activities are best suited for maintaining the quality of each GRS wall layer. Note that in the RSF and the integrated approach, a geotextile must be used to prevent migration of fill material and erosion.

MATERIALS

Once materials are delivered to the site, they should be inspected for compliance with the guidelines and project specifications. Materials should be visually inspected for quality, damage, and defects.

- *Backfill*: In addition to the quarry material certificate showing the gradation of the aggregate, a visual inspection should be performed to verify maximum grain size, amount of fines and grain shape (angular or rounded), excess fines, moisture content, and durability.

- *Facing block*: The facing block should be inspected for integrity, consistency, and dimension tolerances. Confirm that sufficient quantities and proper block type (solid block) corners, and face block, are present onsite and ready for use.
- *Geosynthetic reinforcement*: Verify that the specified type and strength of geosynthetic is correct along with the required roll dimensions. Provides detailed tests that should be documented for each roll of reinforcement.

EQUIPMENT

Compaction of the backfill in a GRS wall or abutment is a critical construction activity. It should be confirmed that the compaction equipment onsite is compatible with the selected backfill material. Verify that the required hand tools are onsite for spreading and grading aggregate, maintaining the facing alignment, and sweeping the top of the CMU facing block.

PROJECT LAYOUT

Verify that all layout reference points are established, with particular emphasis on the location of the following areas:

- Center line of superstructure.
- RSF area within lines and grade of working drawings.
- Bearing area of the bridge beams.
- Wing wall width and length.
- Clear space and setback.
- Span length.
- Center of bearing to center of bearing.
- Elevations.
- Grades.

CONSTRUCTION ACTIVITIES

GRS is built from the bottom to the top. Those responsible

for inspection need to make certain each layer is constructed and tested in accordance with the contract drawings and specifications before proceeding with subsequent layers. Inspection and QC/QA activities are discussed in this section.

- *Working bench*: Before excavation, the working bench/ platform needs to be inspected for stability with consideration for drainage. Any movement should be controlled.
- *Foundation excavation*: The foundation should be cut as outlined in the plan and inspected for any soft areas before compaction and proof rolling.
- *Geotextile–wrapped RSF foundation*: For encapsulation of the RSF, it should be confirmed that the open edge of any overlap is facing downstream and the three sides 'two wing walls and one abutment face wall are contained by a layer of geotextile to prevent erosion.
- *Leveling course*: In order to set the first course of facing block level and plumb, the top elevation of the RSF should be as close to grade as possible. Often, a thin (0.5–inch) leveling layer of aggregate is placed under the first course. Inspection of this leveling layer should be performed to determine its thickness or the need to replace it with a low slump wet concrete/grout mix.
- *Compaction of backfill*: Inspection of backfill operations should verify compliance with the construction guidelines outlined. Compaction behind the wall face and within the bearing area is important. Inspection should confirm that each lift never exceeds the specified thickness.
- Compaction control should be maintained through field density tests or other soil stiffness–based methods. For backfill material containing fines (minus No. 200), the moisture content should be within the specified range (±2 per cent). This improves the compaction process. Compaction of open–graded aggregate should be observed to ensure nonmovement of material under the compaction equipment; this

observation is an indication of compaction or stiffness and is dictated by the number of passes.

- *Reinforcement installation*: Inspect the installation of each reinforcement layer to ensure it is properly placed, has adequate facing element coverage for the frictional connection, and is free of wrinkles. Anticipate the location and placement of the bearing bed reinforcement layers, particularly in situations when the bridge is superelevated.
- *Facing block placement*: Prior to placement of the reinforcement layer over the facing block, the block should be inspected to verify a clean surface. This is essential in maintaining wall alignment and avoiding block cracking due to point loads. The inspection process should ensure that there is no rocking motion when setting the block, which can be indicative of point load bearing.
- *Wall alignment*: Visual inspection should be performed at regular intervals during construction. This will help ensure that the wall is within vertical and horizontal tolerances for alignment. QC should be performed on block alignment using a string line on at least every third course. Vertical alignment can be checked with a plumb bob.
- *Wall termination*: Make sure that all wing wall terminations are sufficiently embedded to prevent undermining from erosion. A terminated course needs to be founded on a stable compacted layer of granular fill material as outlined or on an excavated cut into native soils.
- *Fill slope side*: The fill slope at the wing walls is usually built with native soil as the GRS wall advances upward. The fill slope should be constructed with a drain path that leads away from the wall face. Surface run-off should be diverted to prevent saturation of the soil fill slope. Temporary drainage may need to be installed to preserve the integrity of the cut slope.

- *Site drainage*: The working platform should be compacted and graded to drain surface water away from the working area. Any pit excavation should be sloped to drain to a location that can be pumped.
- *Heavy equipment operation*: It is beneficial to have construction equipment centrally located in the work area and to have materials strategically stocked near the equipment for efficient transfer to the labour crew. Equipment operators should take caution when working near large layers of exposed geosynthetic.
- *Beam seat*: Construction of the beam seat should be inspected to confirm the use of methods described. It is important to verify that the beam seat is constructed at the correct elevation and grade to provide the specified clear space and setback.
- *CMU core grouting*: The core of the top three courses of CMU blocks should be filled with concrete wall mix. Rebar dowels should be cut to length (20 inches) and inserted into the core of the top three courses. The concrete mix should be rodded with a rebar dowel before insertion to eliminate voids. Sufficient concrete should be available to form the coping cap during the same pour.
- *Wrapped integrated approach*: Prior to placement of the geotextile reinforcement, it should be verified that the length is adequate to wrap the fill and extend back towards the road as shown in the one–sheet plan. Sufficient reinforcement width should also be available to laterally confine the approach fill if necessary.

The lift thicknesses for each lift should be checked to ensure that they do not exceed the maximum thickness and that secondary reinforcement is placed within fill layers that are greater than 8 inches. If a granular road base is used in the wrapped approach, verify that its compaction conforms to density requirements for the road as well as the GRS.

At the top of the integrated approach, verify that a 1– to 2–inch layer of aggregate is placed on the top reinforcement layer for protection from hot mix asphalt. Verify that the paving fabric, if used, bridges the interface from the deck to the approach as described.

DOCUMENTATION

COMPLIANCE DOCUMENTATION

Field test results should be carefully measured and archived as a permanent part of the job record. This information can also be used to modify field (construction and inspection, practices).

The main field measurement, moisture density tests, should be documented during construction. Other documentation should include construction modifications, field changes, and daily construction reports.

RECORD DRAWINGS

As–built plans should be prepared and provided to the owner upon completion of the project.

CONTRACTING METHODS

Of the two types of contracting methods commonly used for specialty construction, the procedural method and performance method, the preferred approach for the GRS–IBS is the procedural method. GRS performance–based methods can be developed when the technique becomes more widespread. The generic nature of GRS walls and abutments fits well with the performance–based method, which can advance the technology by creating an opportunity to develop new techniques, details, and equipment.

PERFORMANCE METHOD

In a performance–based contract, the contractor can choose a GRS system based on its performance and constructability. The contractor should verify that the GRS–IBS is constructible and performs as outlined in the requirements. Careful attention should be placed on the compatibility between the backfill material, the wall facing, and the reinforcement to assure that the wall meets the necessary requirements. Under this method, design and performance criteria should be based on the data provided. Material and construction specifications can be based on the information. This contract method requires that the reviewers have considerable knowledge in GRS technology to accept design submittals.

PROCEDURAL METHOD

In this contract method, the agency or owner provides a detailed set of design plans and construction specifications in the bid document. QA begins with an initial plan, design, and review of construction materials. Approval should be dependent on someone experienced in the design and construction of the GRS system. Also, the completed project should be in compliance with local agency building codes and regulations.

Fully detailed plans and items requiring review prior to initiating a GRS project should consist of the following:

- Design calculations.
  - Stability analysis.
  - Bearing capacity.
  - Hydraulic analysis.
  - Loads.
- Project drawings
  - Plan drawings.
  - Cross sectional drawings of all abutment wall faces and wing walls.
  - Elevation drawings.
  - Horizontal and vertical curve details.
  - Construction details addressing guardrails, parapets, beam seat, wing wall configurations, etc.
  - General notes.
  - Fabric schedule.
  - Block schedule.
  - One–sheet plan for quick reference.
- Geotechnical report.
  - Plan view of testing.

  - Subsurface profile (if necessary).
  - Test boring logs (if necessary).
  - Laboratory test data.
  - Engineering properties of foundation soil, retained soil, and settlement analysis.
  - Allowable/ultimate bearing pressure of foundation soil.
  - Ground water and free water conditions.
  - Existing abutment conditions 'if replacement bridge.
  - Historical flood events.
- Hydrology report.
  - Annual Q50, Q100, and Q200 floodwater levels and velocities.
  - Stable particle size analysis for scour potential.
  - Land uses that could impact flood levels.
- *Verification of experience (prequalification)*: The contractor should be certified or prequalified in GRS construction methods in accordance with the procedures presented in this guide. As an alternative, a contractor should be able to verify demonstrated knowledge in constructing GRS structures. For GRS –IBS projects, the contractor should provide information on the successful construction of many GRS walls and abutments. It is important that the contractor have an understanding of the compatible relationship between wall facings and backfill material.
- *QC/QA plan*: A QC/QA plan should be developed by the agency or contractor performing the work and should be followed by the contractor. The QC/QA plan should detail types of measurements and documentation that will be maintained during construction to ensure compliance with GRS guidelines and standards.

CONTRACTOR SUBMITTALS

Materials used to construct GRS–IBS are readily available

from a number of sources. The only requirement is that they meet the standards provided in this manual.

The main materials that should be reviewed prior to construction are as follows:

- CMU block specifications or other type of facing element.
- Backfill gradation, type, and source. The backfill submittal should include aggregates used for the RSF, abutment wall, and integrated approach.
- Geosynthetic reinforcement.

5

Construction

INTRODUCTION

GRS construction uses basic earthwork methods, primarily for excavation and compaction, along with sound general construction practices. The materials are readily available, which is a benefit of the generic nature of the system. Provides guidance on most field–related scenarios. All methods that are presented have been field–tested and applied during the construction of GRS-IBS. The techniques outlined can be applied to efficiently construct the layered system and have been proven to quickly construct the GRS–IBS. The contractor will ultimately choose the methods most efficient for the site, the crew, and the equipment on hand.

The guidance outlined here applies to GRS structures, specifically abutments built with CMU blocks. This guidance can also be adapted to other GRS structures built with different facing systems.

GRS construction has two principal components: (1) logistics and (2) aspects associated with actual construction. Logistics occur after the final design and before construction, outlining a plan for implementation and control of the construction process. Even though building a GRS abutment is as simple as a row of facing block, a layer of well–compacted granular fill, and a sheet of reinforcement, the process will be hampered without adequate planning to ensure optimum flow and placement of material during the course of the project.

As a result, the single–sheet plan was devised to provide information on the reinforcement schedule and the facing block schedule. The single–sheet plan also contains information on the limits of excavation and details about assembly of the GRS structure. A second sheet may be necessary to detail quantities and construction notes.

This chapter conveys the importance of the following details as a means to rapid GRS construction:

- Careful attention to the first row of blocks. Since all other courses of block are built off the first row, it is essential to ensure that the bottom row is level and even for fast construction.
- Optimization of crew size and equipment for enhanced productivity. Too many labourers or excess onsite equipment can cause confusion and slow down the construction process.
- Allowing time for a labour crew to adjust to the construction of the GRS–IBS. Having each crew member do their part in one of three basic steps of GRS construction (laying a course of facing block) compacting a layer of granular backfill, and placing a layer of reinforcement, dramatically improves productivity.
- Establishing the central position of the excavator. Typically, it is best to limit movement of the excavator by locating it towards the back of the abutment, where it can both reach and place material without moving.

LABOUR AND EQUIPMENT

LABOUR REQUIREMENTS

A typical labour crew on GRS–IBS projects has consisted of about five workers: four labourers and an equipment operator. The equipment operator is central to the project and provides support to the labour crew. The equipment operator is responsible for shaping the excavation to facilitate construction of the RSF and the GRS abutment in addition to placing fill material and moving facing units into the work area. Typically, one member of the labour crew has the role of foreman and is responsible for layout of excavation limits, grades, alignment of wall face, placement of facing blocks, compaction of fill, placement of geosynthetic reinforcement, and other activities to streamline production and the flow of material to the job site.

Fig. 5.1 Typical Labour Crew with Centrally Located Track Hoe.

TOOL AND EQUIPMENT REQUIREMENTS

Specialized equipment is not required to construct GRS–IBS. Simple tools that are readily available and relatively inexpensive can be used. These include hand tools, measuring devices, and heavy equipment. The contractor may modify the included lists depending on the site, the crew, and the size of the IBS.

Typical hand tools include the following:

- Gravel rake (concrete spreader).
- Shovels (flat blade and spade).
- Heavy rakes.
- Broom to sweep top of blocks.
- Wisk broom.
- 2– to 3–lb sledgehammer and wood two–by–fours to align blocks.
- Heavy rubber mallet.
- Spade trowel.
- Razor knives or utility knives to cut reinforcement.
- Hand tamper with metal base plate.
- Chainsaw to cut reinforcement roll.
- Concrete saw.

- 5–gallon bucket.
- Block lifter.
- Standard concrete mixing and finishing tools.

Typical measuring devices include the following:

- Survey equipment.
- Laser level.
- String line to align blocks.
- 4–ft carpenter's level.
- Plum bob to check wall batter.
- Measuring tapes.
- Chalk line.

Typical heavy equipment includes the following:

- Walk–behind vibratory plate tampers (200 lb and 18 inches wide or larger).
- Track hoe excavator.
- Riding smooth drum vibratory roller (compacting 3.28 ft from wall face).
- Pallet forks for excavator (for moving CMU block in and out of work area).
- Trash pump and hose for dewatering foundation excavation.
- Backhoe (as needed for material staging).

SITE PREPARATION

GRS is built from the bottom up and generally from within the footprint of the structure. Staging and delivery of materials to the site should allow for continuous GRS construction and effective use of the space.

Fig. 5.2 Cut Slope of Retained Soil.

Delivered material should be easily accessible to the excavator, which is the central piece of equipment. As shown in figure, the excavator is positioned inside the wall area for easy placement of fill, block, and other materials. Labour should be organized to assemble construction materials as needed on the work platform.

SITE LAYOUT

Site preparation begins with a survey of the bridge site to stake limits for the excavation. Reference stakes should be located in an area where they will remain undisturbed during construction of the base of the wall, usually about 5 ft from the excavation. The base of the GRS abutment and wing walls should be constructed to within 1 inch of the staked elevations. The external GRS abutment and wing walls should be constructed to within ±0.5 inches of the surveyed staked dimensions.

EXCAVATION

All excavations should comply with Occupational Safety and Health Administration requirements. Excavation of the site involves shaping the slope for temporary slope stability, safety, and constructability.

The temporary cut in the retained soil should be designed to accommodate movement of labour. The design of a temporary excavation needs to consider the loading imposed by heavy equipment and the reach limits of the excavator. figure shows a typical cut slope in stiff clay. The excavation should include provisions for drainage with a sloped cut to facilitate the movement of water.

Any open excavations that form a pit should be backfilled with crushed aggregate and compacted. Excavation also includes the clearing and grubbing of vegetation. In situations where the retained fill is stable, the volume of excavation can be limited to reduce the size of the GRS mass. In the case of an abutment application, this would form a horseshoe shaped excavation, as shown in fig. 5.3.

Fig. 5.3 Horseshoe–Shaped Excavation with Native Soil Still Intact in Middle.

PLACEMENT OF ABUTMENT BEHIND EXISTING SUBSTRUCTURE

In some situations, it may be beneficial to build GRS–IBS behind an existing substructure. Project feasibility, environmental considerations, and other factors need to be assessed before selecting this type of project layout. Building the bridge behind an existing substructure often requires the removal of the top part of the abutment walls to provide additional space for the width of the new GRS–IBS. figure through illustrate this technique. Note that the design of the GRS-IBS will be the same whether it is built behind an existing abutment or not.

Fig. 5.4 GRS–IBS Built Behind an Existing Concrete Abutment.

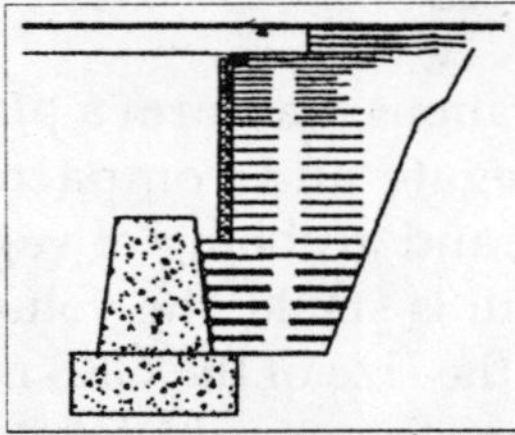

Fig. 5.5 Illustration. Cross Section of GRS–IBS Built Behind an Existing Concrete Abutment.

Fig. 5.6 Building the RSF Behind an Existing Abutment.

RSF

The depth and footprint of the excavation for the RSF should be based on external stability, as described. The base of the RSF should be cut smooth. It should be excavated to uniform depth, and all loose, unstable material should be removed from the site. If the base of the excavation is left open, it should be graded to one end to facilitate the removal of any intrusion of water with a pump. If flooded, all water should be removed along with soft, saturated soils. The excavation should be backfilled as soon as possible to provide a suitable foundation and avoid adverse weather delays. The construction of the RSF can typically be completed in less than one day but is dependent on the size and depth of excavation, type of materials, equipment, and experience.

Fig. 5.7 RSF Excavation below Stream Level.

The base of the excavation should be compacted before construction of the RSF. This may require proof rolling, and any soft spots or voids should be backfilled with compacted fill material. Figure shows the preparation of the RSF cut.

Fig. 5.8 RSF cut Preparation.

The RSF should be encapsulated in geotextile reinforcement placed perpendicular to the abutment face to protect it from possible erosion. The reinforcement sheets should be measured and sized to fully enclose the RSF on three sides: the face and the two wing wall sides. If the GRS abutment is adjacent to water, the reinforcement sheets should overlap, starting with the first layer on the upstream side of the RSF. All overlapped sections of reinforcement in the area of the RSF should be oriented to prevent running water from penetrating the layers of reinforcement.

The first layer of reinforcement should be placed on the upstream side of the abutment with subsequent layers, if needed, overlapped a minimum of 3 ft on the downstream side. This prevents water from infiltrating the RSF. The wrapped corners of the RSF need to be tight and without exposed soil within the RSF to complete the encapsulation.

Fig. 5.9 Encapsulation of Fill in RSF.

Typical reinforcement spacing in the RSF is 12 inches. The reinforcement should be pulled taught to remove all wrinkles prior to placing and compacting the structural backfill. Fill should be placed from the face to the back to roll folds or wrinkles to the free end of the reinforcement layer. The RSF should be constructed with structural fill, as specified. The structural fill is to be compacted in accordance with in compacted lifts not to exceed 6 inches (with two compacted lifts per each 12–inch layer). The first course of wall block sits directly on the RSF, as shown in figure, so it is important that the fill material is graded and level before encapsulating the RSF.

Fig. 5.10 Placement of Wall Block on Wrapped RSF.

While the base of a typical GRS abutment is built with solid CMU, damage can occur during the placement of channel rock protection or from other large pieces of concrete rubble that extend above the solid block zone. Riprap protection should be placed in a manner to prevent damage to the CMU wall face. Impact of large rock or concrete fragments during placement can crack the CMU block. Larger rocks should be uniformly distributed and placed firmly in contact with each other, with smaller rocks and fragments filling the voids between the larger rocks. This procedure often requires hand placement of smaller rocks to fill the voids. If any CMU block is damaged, refer to repair procedures.

COMPACTION

Compaction of the backfill should be to at least 95 per cent of maximum dry density according to AASHTO T–99. Backfill material containing fines should be compacted at a moisture content close to optimum (±2 per cent). Lifts of 8 inches should be compacted using vibratory roller compactionequipment. The facing blocks provide a form for each lift of fill. Other stiffness–based compaction control methods can be used. For open–graded fills, compact to non–movement or no appreciable displacement and assess with visual inspection.

Since the facing elements are not rigidly connected to the reinforcement, hand–operated compaction equipment is required within 1.5 ft of the front of the wall face. It is very important for adequate GRS performance that the backfill is properly compacted. The top 5 ft of the abutment should be compacted to 100 per cent of the maximum density according to AASHTO T–99. Onsite compaction equipment should be selected to achieve the required density of the fill materials. Considering that compaction is critical to the success of the project, compaction equipment should be in good operating order for efficient use. In addition, backup equipment should be available to provide quality construction throughout the project and to avoid construction delays.

COMPACTION PROCEDURE

Once fill is placed at the required thickness and graded, all areas behind the CMU block should be compacted to the required density. Any depression behind the facing block should be filled level to the top of the CMU block prior to compaction.

Compaction directly behind the CMU block should be performed in a manner that maintains wall alignment while improving the density of fill behind the block.

This can be achieved in the following ways:

- Placing a fill lift directly behind the CMU block face and rodding or foot tamping along the row of CMU block while exerting downward pressure on the block to prevent lateral movement. For multiple lifts, the top lift height is slightly higher than the block to compensate for compression of the fill during compaction.
- Using a lightweight vibratory plate compactor directly behind the CMU block while exerting downward pressure on the block to prevent lateral movement.

- Using larger vibratory compactors for the remainder of the fill area 3 ft from the face of the GRS wall. Check for outward block movement and adjust accordingly.

The most common compaction QC tool is the nuclear density gauge. Other instruments are also available for compaction control such as the Clegg hammer, the soil stiffness gauge, or the falling weight deflectometer. These devices are typically used by correlating their measurements to soil density and moisture content. Method–based compaction specifications can also be used. For open–graded fills, compact to non–movement or no appreciable displacement and assess with visual inspection.

REINFORCEMENT

Generally, the length of the reinforcement layers will follow the cut slope, as shown in figure. While the reinforcement layers in the GRS abutment can be any geosynthetic, the RSF and integrated approach should be constructed and encapsulated with a geotextile to confine the compacted granular fill. The geosynthetic should be placed so that the strongest direction is perpendicular to the abutment face, as shown in figure. Where the roll ends, the next roll should begin. Overlapping between sheets is required. The geosynthetic reinforcement should extend between layers of CMU block to provide a frictional connection. The geosynthetic reinforcement should cover a minimum of 85 per cent of the top surface of the CMU block; any excess can be removed by either burning with a propane torch or cutting with a razor knife.

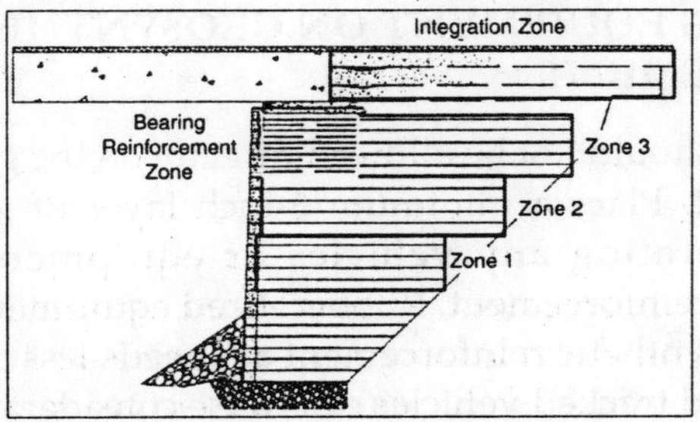

Fig. 5.11 Illustration. Typical Reinforcement Zones.

Fig. 5.12 Reinforcement Rolled out Parallel to Wall Face.

After the geosynthetic is rolled out, it should be laid so that it is taut, free of wrinkles, and flat. The geosynthetic can be held in place with the fill. Placement of fill should be from the wall face backward to remove and prevent the formation of wrinkles in the geosynthetic. A conscious effort should be taken during placement of fill to prevent the development of wrinkles.

Splices of reinforcement can occur without overlap. Splice seams should be staggered to avoid a continuous break in the reinforcement throughout the GRS structure. All splice seams should run perpendicular to the wall face. Overlaps of adjacent geosynthetic should be trimmed where they are in contact with the surface of the CMU block to avoid varying geosynthetic thicknesses between the CMU block. Any seams in the geosynthetic should be staggered with each successive layer of the GRS abutment. All seams between adjacent sheets of geosynthetic located in the area beneath the footprint of the bridge seat should be perpendicular to the abutment wall face.

OPERATING EQUIPMENT ON GEOSYNTHETIC REINFORCEMENT

Driving should not be allowed directly on the geosynthetic reinforcement. Place a minimum 6-inch layer of granular fill prior to operating any vehicles or equipment over the geosynthetic reinforcement. Rubber–tired equipment may pass over the geosynthetic reinforcement at speeds less than 5 mi/h. Skid steers and tracked vehicles can cause considerable damage to the geosynthetic. On one occasion, a track hoe operating on a GRS structure turned and pulled the fabric causing deformation to the wall face.

For this reason, it is recommended to restrict the use of these vehicles on GRS structures. If absolutely necessary, use may be permitted provided no sudden braking or sharp turning occur and a minimum 6–inch cover is placed.

BEARING REINFORCEMENT BED

The bearing reinforcement bed provides additional strength in the upper GRS wall layers directly beneath the bearing area of the superstructure. These reinforcement layers are not sandwiched between two consecutive rows of block but are placed behind the CMU block at 4–inch spacing. This 4–inch reinforcement spacing is generally placed in the top five layers of the GRS abutment or as determined by design.

Bearing bed reinforcement spacing in superelevated abutment walls requires additional planning. The 4–inch reinforcement spacing needs to be in place for the top five courses of block at the lowest elevation across the abutment wall. The reinforcement schedule will guide field personnel in the proper placement of the geosynthetic along a wall block course.

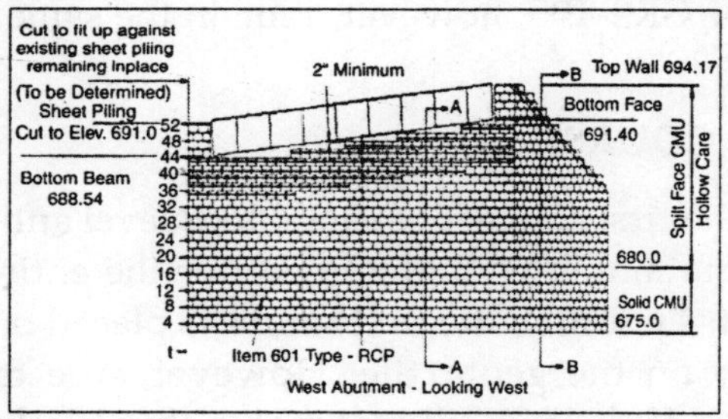

Fig. 5.13 Illustration. Superelevation Reinforcement Schedule.

SUPERELEVATION

The reinforcement layers become stair–stepped in the upper wall layers as the superelevation of the abutment is constructed. The reinforcement terminates along the angle surface of the superelevation. The GRS wall reinforcement schedule should show the termination of each layer of reinforcement across the abutment wall from low to high elevation.

Fig. 5.14 Superelevation Reinforcement Layers.

WALL FACE

This manual focuses on the use of CMU for the wall facing; however since GRS is internally stable, any facing elements can be used in construction.

For flexible facings other than the recommended CMU block 'including wrapped, timber, natural rock, or welded wire basket formed facing,, other construction guidelines may need to be followed. These are outlined by Wu *et al*. The design guidelines for GRS–IBS, however, remain the same as those in this manual.

LEVELING COURSE

Setting the first course of facing block level and to grade is critical in maintaining wall alignment for the entire height of the abutment. Typically, the first course is placed on top of the RSF, directly on the geotextile. However, due to the large aggregate size of the RSF fill material, a thin leveling layer of fine aggregate can help set the CMU blocks to grade and prevent them from rocking.

The leveling layer should be kept to a minimum thickness, no more than 0.5 inches. If the leveling layer exceeds this thickness and there is the potential for water to erode and undermine the aggregate, mortar or grout should be placed in the gap between the RSF and the first course.

SETTING THE CMU BLOCK

CMU block wall construction should begin at the lowest portion of the excavation with each layer placed horizontally, as shown on the plans. Each layer should be constructed entirely before beginning the next layer. A stretcher or running bond should be maintained between courses of block so that the joints between the blocks are offset with each row. If a designed scour countermeasure such as a riprap apron is used, a geotextile filter fabric should be placed under the apron and anchored between the first and second courses of CMU block.

Since the CMU blocks are dry stacked without mortar, it is important to avoid cracking the block and to maintain a horizontal uniform elevation by sweeping the top surface of the block clean of debris and fill material prior to the placement of the next layer of geosynthetic and CMU block. Gravel material between layers of block creates point loads that can cause cracks. Also, gravel material between the blocks causes them to rock, making it difficult to secure a good fit.

When setting a course of block, each block should be placed tightly against the adjoining block, preventing gaps from which fill material can escape. Before proceeding to the next layer, it is often useful to walk along the top of the blocks to easily identify a poorly seated block.

In order to avoid cutting a block when the CMU block schedule shows the wall terminating with half a block, a full CMU block can be turned 90 degrees, placing the 8–inch width towards the face. This typically occurs at the termination of a wing wall. The end block that forms the termination does not have to be a corner CMU (with two finished sides) because the ends of most wing walls are embedded into the fill slope.

WALL FACE ALIGNMENT

When placing and compacting fill behind the CMU block, it is sometimes necessary to set the block back about 0.5 inches to allow for lateral outward movement of the CMU block during compaction. It should be noted that each combination of wall facing and backfill reacts differently during the compaction process, and adjustment of the setback distance between block courses should be performed as needed to maintain the necessary batter.

The vertical GRS wall should be checked for plumbness at least every other layer, and any deviations greater than 0.25 inches should be corrected. Before placement of the backfill, every other row of block alignment should be checked with a string line referenced off the back of the facing block from wall corner to corner.

Fig. 5.15 Checking Block Alignment with String Line Reference from Back of Block.

If CMU blocks become displaced during construction, they can often be hammered back into position using a 3–lb sledgehammer and a block of wood as protection. If the CMU block is excessively out of alignment, the fill material needs to be excavated, the CMU block repositioned, and the fill material replaced and recompacted.

BLOCK ALIGNMENT FOR BATTERED WALLS

Block alignment for battered walls is similar to that for vertical walls. In abutment situations where the face wall turns to form the wing wall, however, it is necessary to trim blocks on either end to account for the reduced wall length. All cuts should be performed to maintain the standard running or stretcher bond between the rows of dry–stacked blocks, with the vertical joints of each course midway between those of adjoining courses. In special situations, negative battered walls have been constructed when the top area needs to be greater than the bottom, as in the case of road widening shown in fig. 5.16. The negative batter can be created by offsetting the CMU block by a measured amount in consecutive wall layers then filling and compacting as specified.

Fig. 5.16 Negative Batter Wall Face.

SUPERELEVATION

When the plan shows a superelevation for the bridge, the top courses of CMU beneath the superstructure should be trimmed to match the elevation difference and clear space across the abutment.

This will produce a sloped face wall and aid in construction of the beam seat. One method is to snap a chalk line along the back of the block at the superelevation slope. A carpenter's angle finder can also be used to mark the cut.

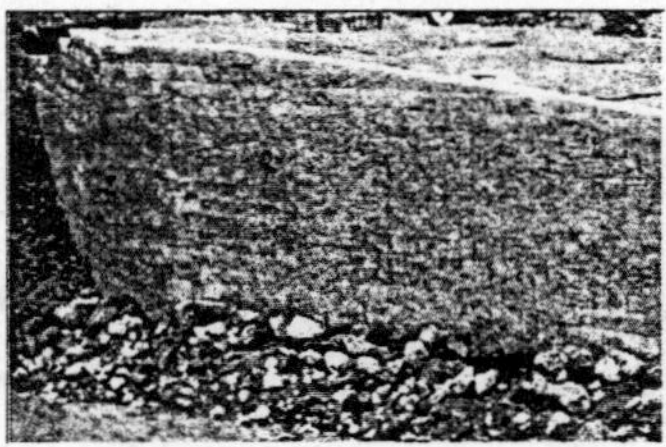

Fig. 5.17 Blocks Trimmed to Match Superelevaticn.

WALL CORNERS

Right–angle wall corners, as shown in fig. 5.18, are

constructed with CMU corner blocks that have architectural detail on two sides, providing an aesthetic finish. Facing wall and wing wall courses should be staggered to form a tight, interlocking, stable corner.

Fig. 5.18 Right–angle Wall Corner.

Walls with angles larger or smaller than 90° require additional effort. The corner blocks need to be cut to form the angled face. As a result, a vertical seam or joint is formed at the corner. Corners with vertical seams may have open block joints, making it prudent to fill the corner blocks with a concrete mix and install bent rebar to close and connect the seam at each course of block, as shown in fig. 5.19. This procedure secures the two faces and prevents compaction–induced separation during construction of subsequent GRS layers. It may also be used wherever added strength at the wall corner is desired.

Fig. 5.19 Vertical Seam in Wing Wall.

Fig. 5.20 Rebar Installed in Vertical Seam Prior to Grout.

The top three courses of CMU block in the abutment are susceptible to movement simply from not having the weight of successive layers holding them in place. To prevent displacement, the hollow cores of the top three courses of CMU blocks are filled with a concrete wall fill and pinned together with No. 4 rebar, preferably epoxy–coated and embedded with a minimum 2–inch cover.

Fig. 5.21 Connecting the Top Courses of Blocks.

To grout and pin the top of the wall, the reinforcement between the top two courses of CMU block needs to be removed to open the core for placement of concrete wall fill and a 20–inch–long No. 4 rebar dowel, preferably epoxy–coated with 2–inch cover. This can be accomplished either by cutting the reinforcement with a razor knife or by burning the reinforcement.

The concrete wall fill is placed in two steps. After the block void is filled with concrete to the top of the block and the steel rebar is inserted, a thin layer of the same concrete mix is placed on top of the block to form the coping cap, as shown in fig. 5.22.

The coping is then hand–troweled either square or round and sloped to drain. A wet–cast cap is more durable than a dry–cast cap and eliminates the need to furnish and install a separate cap unit.

Fig. 5.22 Rounded Coping Cap.

Fig. 5.23 Square Coping Cap.

Once the top of wall has been grouted and pinned, care should be taken to avoid any construction activity that may pull on the top layer of reinforcement. The frictional connection between the block is strong, and when courses are pinned together, the entire grouted wall face can be pulled out of alignment.

BEAM SEAT

The beam seat is constructed directly above the bearing bed reinforcement zone. The superstructure is then positioned on top of the beam seat, as shown in fig. 5.24. The purpose of the beam seat is to ensure that the superstructure bears on the GRS abutment and not the wall facing block and to provide the necessary clear space between the superstructure and the wall face. Typically, the clear space is 3 inches, or 2 per cent of the abutment height, depending on the required design.

Fig. 5.24 Box Beam Placed on Beam Seat.

Fig. 5.25 Detail View of Box Beam Placed on Beam Seat.

In general, the thickness of the beam seat is approximately 8 inches and consists of two 4–inch lifts of wrapped–face GRS. Remember, before construction of the beam seat, the cores of the CMU blocks on the abutment wall face must be pinned with No. 4 rebar and filled with concrete wall mix.

Fig. 5.26 Bearing Area Block Grouted Prior to Beam Placement.

BEAM SEAT PROCEDURE

Once the block elevation beneath the bearing area is established and the hollow cores are filled with grout, the beam seat is ready for construction.

The following steps should be used:

- Place precut 4–inch–thick foam board on the top of the bearing bed reinforcement. Sometimes, a thin layer of backfill may be necessary beneath the foam board for grading purposes and to ensure the proper clear space height and drainage 'crown in bridge. The foam board should butt against the back face of the CMU block. The exposed edge of the foam board helps form the nose of the reinforcement wrap across the length of the bearing area.

Fig. 5.27 Foam Board and 4–inch Block Assembly to form Beam Seat.

- Set 4–inch solid concrete blocks on top of the foam board across the entire length of the bearing area. The back edge of the top CMU face block holds the 4–inch concrete block in place during compaction. Note that the distance between the top of the grouted CMU block and the top of the beam seat 'the clear space, is the distance the beams can settle before bearing on the facing blocks.

Fig. 5.28 Inch Concrete Block on top of foam Board Against top CMU Face Block.

- Use the first 4–inch wrapped layer of compacted fill as the thickness to the top of the foam board.

Fig. 5.29 First 4–inch Wrap Butted Against foam Board.

- Place the second 4–inch wrapped layer of compacted fill to the top of the 4–inch solid block, creating the clear space as shown in fig. 5.30. The top of this layer controls the beam elevation, and should therefore be carefully compacted and graded.

Fig. 5.30 Top 4–inch Wrap Butted Against 4–inch Solid Block.

- Before folding the final wrap, it may be necessary to grade the surface aggregate of the beam seat slightly high, to about 0.5 inches, to aid in seating the superstructure and to maximize contact with the bearing area.

SETBACK

The setback is the distance between the back of the facing block and the front of the beam seat. This distance can be established during construction of the beam seat and placement of the block and foam board used to form the beam seat wrap. The setback distance is usually 8 inches but can be greater.

ALUMINUM FLASHING

The aluminum flashing drip edge is installed prior to setting the bridge beams and is placed in between the bottom of the beams and the foam board. The flashing is held in place by the pressure of the beams on the compressible foam board. The length of the flashing should extend beyond the outside edge of the bridge beams and be trimmed to fit against the parapets.

Fig. 5.31 Aluminum Flashing (Drip Edge) between Beams and Top of CMU Block.

CIP OR PRECAST FOOTING

For GRS–IBS built without adjacent concrete beams, a CIP or precast footing may be necessary, as with steel beams or spread girders.

Fig. 5.32 Steel Girder on CIP Footing.

PLACEMENT OF SUPERSTRUCTURE

CRANE POSITION ON GRS MASS

The crane used for placement of the superstructure can be

positioned on the GRS abutment provided the outrigger pads are sized within the capacity of the GRS mass. The outrigger pads should be sized for 4,000 psf near the face of the abutment wall with greater loads able to be supported with increasing distance from the abutment face.

Fig.. 5.33 Outrigger Pads Near Wall Face.

REINFORCEMENT OF BEAM SEAT

An additional layer of reinforcement should be placed between the beam seat and concrete or steel beams to provide additional protection of the beam seat. The additional layer of reinforcement may decrease the sliding resistance between the superstructure and the beam seat.

Fig. 5.34 Additional Reinforcement Under Beam.

SETTING SUPERSTRUCTURE ON BEAM SEAT 'WITHOUT CIP FOOTING

Since the bearing surface is aggregate under a layer of geosynthetic reinforcement, it is important to set beams square and level. They should never be dragged over the beam seat surface, which could create the potential for an uneven bearing area or a void under the beam, producing uneven bearing stresses between bridge elements.

WING WALLS AND PARAPETS

Wing walls and parapets are constructed after the superstructure is set. The CMU block in the parapet wall should be trimmed or saw cut for a custom fit against the beam edge to prevent the loss of fill material. Figure 5.35 show the construction of the parapet against the superstructure. If the gap between the superstructure and the facing block is difficult to fill using thin slices of cut block, a mortar mix should be used to close the space.

Fig. 5.35 Parapet and Wing Wall Construction, View 1.

Fig. 5.36 Parapet and Wing Wall Construction, view 2.

APPROACH INTEGRATION

Proper approach construction at the road and superstructure interface is essential to minimizing settlement in front of the bridge beams and eliminating the bump at the end of the bridge. This is accomplished by compacting and reinforcing the approach fill in wrapped geotextile layers and blending the integration zone with the approach road base course. The material for the integration zone should be well–graded, as outlined.

Once the superstructure is in place, the approach to the bridge can be constructed using the following steps:

- Trim a geotextile reinforcement sheet to provide the planned length after it is wrapped, and place it behind the beam end. The width of the sheet should allow for wrapping of the sides after the fill layer is placed and compacted. Wrapping of the sides prevents lateral migration of the fill.

Fig.5.37 Placing Reinforcement.

- Place a 6–inch lift of fill and compact, per compaction specifications for road base. Add a secondary layer of reinforcement on top of the 6–inch lift, and then place another 6–inch lift of fill and compact. Fold back the reinforcement sheet to wrap the compacted fill layer and smooth wrinkles.

Fig. 5.38 First 6–inch Fill Lift.

Fig.5.39 Secondary Reinforcement Sheet.

Fig. 5.40 Completed Wrapped Approach Layer.

- Repeat these steps until approximately 2 inches from top of beam grade, as shown in figure 5.41.

Fig. 5.41 Second 6–inch Fill Lift.

Multiple sheets can be used along the width of the approach, as long as all seams are kept perpendicular to the beam ends. The typical wrap reinforcement spacing is 12 inches, with intermediate layers spaced at 6 inches and compacted in 6–inch lifts. However, in the case of beams with a reduced depth, the spacing of the wrapped layers may need to be reduced and the intermediate layers eliminated. The top wrap fold should increase in length with each successive wrapped layer until the fill is 2 inches below the bridge grade.

WRAPPED REINFORCEMENT LAYERS ON SIDES

If lateral spreading of the fill in the integrated approach will be an issue (wing walls are not sufficient to confine the fill at the sides), the reinforcement sheets comprising the wrapped layers should be folded over along the sides and perpendicular to the bridge.

Fig. 5.42 Completed Approach Fill.

PRELOADING

In some situations, it might be beneficial to preload the abutment before paving to minimize postconstruction deformation or settlement within the GRS mass. A simple method of preloading can be achieved be parking fully loaded trucks on the bridge for several days before placement of the asphalt pavement.

PAVING

The top layer of reinforcement should be kept approximately 2 inches below the beam grade. This will allow a layer of aggregate cover to be placed to protect the reinforcement from contact with hot mix asphalt. When GRS–IBS is built with adjacent precast concrete beams, a layer of paving fabric is extended over the beams onto the approach way. Extending the paving fabric 3 ft over the beam approach interface is recommended. This is necessary to bridge the gap and provide an interface to accommodate thermal movement, minimize surface water infiltration, and prevent cracks in the road. Note that paving fabric is already used on top of the beams as a barrier to water infiltration and to absorb stresses to minimize reflective and fatigue cracking of the new asphalt surface layer.

GUARDRAIL POST

Steel H posts are recommended for any railing that is driven through the reinforcement. It is also possible to drill through the GRS mass with an auger to set other types of posts.

SITE DRAINAGE

The GRS–IBS construction area should be protected from surface run-off during the project. Critical areas are behind the abutment wall at the interface between the GRS abutment and the retained fill, at the base of the abutment, and at any location where a fill slope meets the wall face. Design needs to include provisions for surface drainage along the fill slope adjacent to the wing walls. Provisions for drainage should also be included at the boundary of the wing walls and the fill slope. Long walls built along variable elevation or abutment wing walls are often stepped to reduce excavation. In these situations, the termination of wall steps should be sufficiently embedded to prevent problems with erosion. The drainage swell or channel should be separated from the wall to avoid flow directly against the wall face.

Site preparation for drainage should include the following:

- *Grading*: The site should be graded every night in anticipation of precipitation to avoid saturation of soil.
- *Diversion trenches*: An alternative to grading is placing diversion trenches around the perimeter to divert water.
- *Compaction of loose soil*: Any loose soil placed to construct GRS should be graded and compacted before stoppage of work for the day. Also, onsite stockpiles of fill material containing fines should be protected from excess precipitation.

6

Service Performance

INTRODUCTION

The presents methods of evaluating the performance of GRS–IBS, including deformations, thermal movements, and scour monitoring. The performance of in–service GRS–IBS structures can be found in the synthesis report. A distinctive feature in the design of GRS-IBS is that it works with settlement instead of resisting it to create a compatible connection between the approach and the road, providing a long–term solution to the bump at the end of the bridge. Reducing the bump at the end of bridge will improve the overall performance and serviceability of the bridge. This bump not only creates a chronic maintenance issue but also induces an amplification of LL on the superstructure, creating fatigue on bridge elements.

Recently, some owners have shifted their focus on the combined effects of corrosion and fatigue in the evaluation of a bridge's health. The design objective of IBS addresses these two critical durabilityissues by attempting to create a smooth, jointless, affordable bridge system. GRS–IBS was developed to meet the demand for the next generation of small, single span bridges in the United States as part of the FHWA's Bridge of the Future initiative and is built without many of the standard abutment components associated with a traditional bridge (approach slab) sleeper slab, traditional bridge bearings, joint details. The performance of the first series of GRS–IBS indicates that this method has considerable potential to advance the state of the practice.

DEFORMATIONS

The performance of nearly 20 bridges built with GRS–IBS has been an improvement on similar bridges built with conventional construction techniques. The GRS–IBS bridges have performed as well as the conventional bridges structurally and functionally in addition to eliminating the bump at the end of the bridge that often results from conventional construction. The suppression of the bump at the end of the bridge has been maintained for all GRS–IBS bridges that are in service. The first bridge constructed with the IBS method, the Bowman Road Bridge, has been in service since 2005. As of 2010, there has been no development of a crack in the asphalt layer from the road to the bridge.

The total settlement and deformation 'and thus vertical strain, of the GRS abutment due to bridge load is recorded using either a standard survey level and rod system, as shown in fig. 6.1, or an electronic distance measurement (EDM) survey referenced off a permanent survey pole and benchmarks. The precision of all survey measurements 'both the survey level method and the EDM system, is ±0.005 ft.

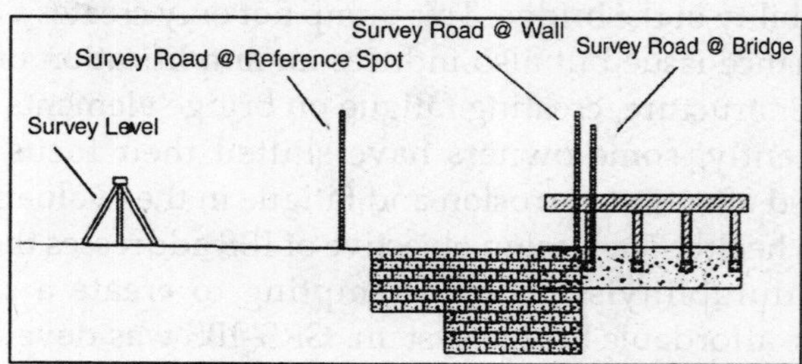

Fig. 6.1 Illustration. Survey Level Method for Superstructure and Wall Settlement.

Using either settlement measurement technique, the settlement is recorded for both the abutment face wall and the superstructure. The difference between the settlement measured on the abutment face wall and the superstructure is the vertical deformation within the GRS mass alone due to the bridge load. To determine secondary settlement, plot settlement versus log–time.

For surveys where a survey level is used, bridge settlement should be measured at four locations (each corner of the bridge) to check for angular distortion and differential settlement. Wall settlement is recorded by the rod off the top of the CMU facing block adjacent to the superstructure, and superstructure settlement is measured with the rod off a guardrail hanger bolt.

For surveys with an EDM system, the total station is referenced off of a permanent pole embedded beneath the frost line and within accessible sight to both abutment face walls. Targets placed on the abutment wall face and the bridge beam or footing are then used to measure movement relative to the permanent pole. Fig. 6.2 shows an example of this for the Tiffin River Bridge in Defiance County, OH. Lateral and vertical movement of the abutment wall face was measured using custom reflective survey targets, which are shown in the black circles in fig. 6.2. Movement of the GRS abutment was measured using targets placed on the concrete footing itself, which are shown in the red circles in fig. 6.2. The difference between the two readings (movement of the abutment and movement of the wall face) provides the compression of the GRS abutment alone, not including foundation settlement. Note that the permanent pole should be installed prior to placement of the steel girders on the CIP footings.

Fig. 6.2 Location of Survey Targets on Tiffin River Bridge.

Fig. 6.3 Location of Total Station Reference Pole.

THERMAL CYCLES

Thermal cycles occur on every bridge structure due to sustained temperature variations. The severity of the expansion and contraction depends on the coefficient of thermal expansion of the bridge. However, observations of bridges built in moderate climates indicate considerable compatibility between the superstructure and the integrated approach, resulting in a smooth transition.

GRS–IBS accommodates movement through the integrated transition zone behind the beam ends. The road base is wrapped with geotextile and then compacted directly against the beam end. This process is described in detail. The wrapped face confines the soil and allows the beam to contract without the fill behind the beam ends sloughing off to fill the void. Because of this, excess pressures behind the beam during expansion are also avoided. The road base is not only wrapped vertically but also laterally to prevent lateral spread.

MONITORING FOR SCOUR

The riprap protection should be monitored during each bridge inspection or after an extreme flood. Any movement of rock should be noted and replaced to prevent scour from progressing and undermining the RSF or the abutment. In all current installations of GRS abutments, no problems have been reported. It should be noted that these installations have been built in non–scour–critical environments and with appropriate countermeasures, as recommend in current practice.

An indicator of scour on an abutment face or wing wall can be achieved by using coloured blocks on the bottom five to eight rows of the abutment. Solid blocks are recommended in the bottom rows as they are more likely to resist any impact of moving riprap, ice, or other abrasion associated with the normal water elevation. The solid coloured blocks are also covered from view by the riprap. Any exposure of coloured blocks could indicate movement or undermining of the riprap, requiring inspection and possibly remediation or repair to protect the RSF and abutment from scour.

SPECIAL REQUIREMENTS FOR HYDRAULIC AND SEISMIC CONDITIONS

HYDRAULIC DESIGN

When bridges are constructed to span a waterway, their foundations must be designed, detailed, and constructed in compliance with section 2.6 (Hydrology and Hydraulics) of the AASHTO *LRFD Bridge Design Specifications* or an FHWA Division Office–approved drainage or bridge manual. These provisions apply equally to both shallow and deep foundations.

GRS–IBS has been successfully used to build abutments near rivers and streams. However, assessing the potential impact of stream instability, scour, and adverse flow conditions is a vital consideration in the decision to use this technology. The potential for issues with stream instability, scour, and adverse flow conditions can lead to deep foundation bottom elevations or expensive countermeasures that could reduce the cost–effectiveness of GRS–IBS abutments. If the potential for abutment scour, contraction scour, long–term degradation, or channel migration is high, costly design considerations or countermeasures could be required.

Other factors, such as channel instability and adverse flow conditions 'skewed approach flow, highly contracted flow, high velocity flow through the bridge opening, etc., at the bridge, could also result in costly design considerations or countermeasures to stabilize the channel against further instability. Any of these conditions might make it advisable to select an alternative bridge abutment technology.

A thorough hydraulic analysis, scour evaluation, and assessment of channel stability of a bridge design will include an appropriate estimate of the design flow, development of water surface profiles through the proposed opening, assessment of scour (abutment) contraction, and long–term degradation,, and if necessary, the design of countermeasures to protect the bridge or stabilize the channel. FHWA and others have developed procedures to assist the engineer in performing these analyses, and these procedures should be followed for GRS–IBS design.

HYDRAULIC DESIGN CONSIDERATIONS

There are a number of important factors to consider when completing a thorough hydraulic design and scour evaluation of a bridge. The determination of a scour elevation based on the computed scour depth; the selection, design and installation of a scour countermeasure; and postconstruction inspection are important factors that must be adequately addressed. The following factors should be considered:

Scour Depth

The scour depth at an abutment is to be calculated as the sum of the depth of contraction scour and long–term degradation. The elevation of the design scour depth is to be calculated by projecting the elevation of the depth of scour from the lowest point in the channel to each of the abutments.

Scour countermeasures

When scour depth is calculated as described in this section, a designed scour countermeasure is included. Design scour countermeasures include riprap aprons, gabion mattresses, and articulating concrete blocks. The purpose of installing a designed scour countermeasure is to prevent loss of soil from underneath a GRS abutment from scour that occurs at or near the abutment. Soil loss can reduce bearing capacity or lead to settlement, which can cause structural failure Fig. 6.4 shows a cross section of a typical abutment riprap countermeasure recommended for smaller, more culvertlike structures (flow length through structure is longer than structure width).

Larger, more bridgelike structures (opening length is greater than the flow distance through the structure) must be evaluated for scour using the procedures outlined in HEC–18 and HEC–20 and use a designed countermeasure as outlined in HEC–23.

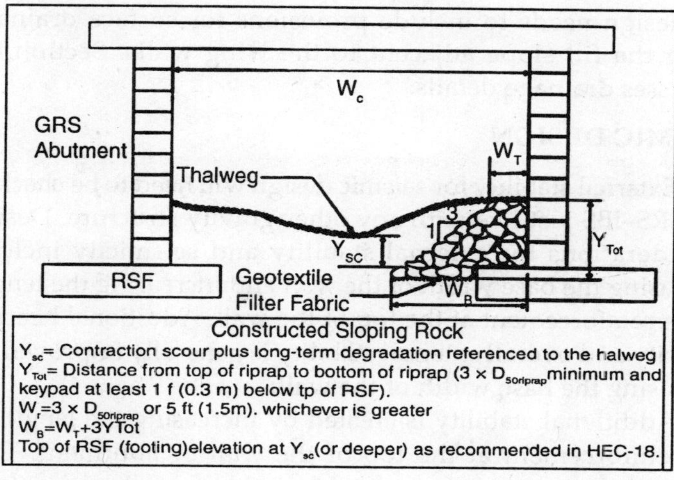

Fig. 6.4 Illustration. Typical Cross Section for Sloping Rock Adapted.

Inspection

After construction, scour countermeasure condition and channel instability should be assessed during each regular bridge inspection and after extreme flood events. Any countermeasure failure or significant change in channel condition should be noted and scheduled for repair or stabilization. Without proper inspection and maintenance, a scour countermeasure may fail or a channel may become unstable, which can lead to undermining of an abutment.

The FHWA's HEC–20 discusses approaches for evaluating channel instability, and HEC–23 discusses approaches for inspection and monitoring the effects of scour. Another hydraulic consideration is drainage. The potential for unbalanced water pressure exists when a wall can become partially submerged by a flood or when surface drainage is not controlled. All GRS structures should include consideration for surface and subsurface drainage.

Critical areas are behind the wall at the interface between the GRS mass and the retained fill, at the base of the wall, and any location where a fill slope meets the wall face. For example, the design needs to include provisions for surface drainage along the fill slope adjacent to the wing walls. Section 6.5 discusses drainage details.

SEISMIC DESIGN

External stability for seismic design will need to be checked for GRS–IBS just like with any other gravity structure. Design considerations for external stability and seismicity include increasing the base width of the wall and increasing the length of the reinforcement at the top of the wall. Additional bearing capacity and overall external stability is generally improved by increasing the base width of the wall.

Additional stability is created by increasing the length of the reinforcement at the top of the wall or abutment. This integrated approach has also been shown to be beneficial because it keys the structure into the existing terrain, preventing the development of a failure plane along the cut slope, which can lead to progressive failure. No seismic design requirements are necessary for the internal stability of GRS–IBS.

Reinforced soil walls have been known to perform better than conventional retaining walls under seismic loading, as evidenced by observations of actual performance in strong earthquake events. A National Cooperative Highway Research Programme (NCHRP) study is being conducted to establish guidelines for the design and construction of GRS abutments under seismic loading. As part of the NCHRP study, a 12–ft–

high GRS abutment supporting a bridge load of about 1,000 kipswas subject to sinusoidal motions on a shake table.

No significant damage or movement was recorded until the acceleration was increased to 1.0 g, at which time base sliding between the GRS abutment and foundation soil became apparent. The superstructure would not have failed due to the deformation of the GRS mass at the 1.0–g acceleration. This experiment suggests that a GRS abutment is capable of withstanding at least low to medium earthquakes without any special provisions.

IMPACT EVENTS

There is limited information on vehicle impact against GRS–IBS. Typically, GRS walls along roadways are built behind a crash barrier. A niche function of GRS technology, however, is rock–fall protection. That application is not covered in this manual, but it serves to show that GRS is capable of withstanding considerable lateral and vertical impacts without failure or loss of serviceability.

7

Design Methodology for GRS–IBS

OVERVIEW OF GRS–IBS DESIGN METHOD

During the past 30 years, GRS technology has been used to build walls, shallow foundations, culverts, bridge abutments, and rock fall barriers. The technology also has been used to stabilize slopes and repair roadways. The focuses on the GRS design method used for GRS–IBS including an abutment and wing walls. While GRS technology can provide solutions in a variety of applications and under certain extreme conditions, the design method described in this manual provides a recipe for design of GRS–IBS with limitations on abutment heights, bridge spans, and design loads.

The design methods described in this appropriate for GRS structures (an abutment and wing walls) with a vertical or near vertical face and at a height that does not exceed 30 ft. Although the majority of bridges built with GRS–IBS have spans of less than 100 ft, spans of up to 140 ft have been constructed. While larger spans are possible, the bearing stress on the GRS abutment is limited to 4,000 lb/ft^2. The demands of longer spans on GRS–IBS are not fully understood at this time, and it is recommended that engineers limit bridge spans to approximately 140 ft until further research has been completed. GRS–IBS abutment capacities are dependent on a combination of the strength of the fill material and the strength of the reinforcement when built in accordance with the two rules of GRS construction: (1) good compaction (95 per cent of maximum dry unit weight) according to AASHTO T99, of high–quality granular fill and (2) closely spaced layers of reinforcement (12 inches or less). It is recommended that design or allowable bearing pressure be limited to 4,000 lb/ft^2.

For design pressures larger than 4,000 lb/ft^2, the performance criteria must be checked against the applicable stress–strain curve resulting from a performance test. The performance criteria for GRS–IBS consist of a tolerable vertical strain of 0.5 per cent and lateral strain of 1 per cent. A significant amount of research and practical experience has shown that GRS–IBS designed and constructed within the limits defined in this manual will produce safe, durable systems.

The design process starts with establishing the project requirements from which the preliminary geometry of GRS–IBS is determined. Once the geometry is defined, it is then evaluated against external and internal modes of failure. An iterative process is used to assess the geometry and make adjustments as necessary to facilitate construction and assure long–term performance. Economy should also be a consideration when evaluating each design alternative.

A general and identifying feature of the GRS–IBS design is a mass built with alternating layers of compacted granular fill material and closely spaced reinforcement (less than or equal to 12 inches). In nearly all of the GRS masses built in the United States as full–scale experiments or as in–service structures, however, the design has been based on an 8–inch layered system. There are other features and principles common to a GRS mass. Most GRS walls have been built with dry–stacked concrete facing blocks and are flexible (in terms of global bending stiffness).

A GRS abutment is a type of gravity structure. Therefore, external stability should be evaluated for the direct sliding, bearing capacity, global stability, and overturning failure modes limiting this type of construction. However, because a GRS mass is relatively ductile and free of tensile strength, overturning about the toe, in a strict sense, is not a possible response to earth pressures at the back of the mass or loading on its top. Other attributes of GRS–IBS also tend to preclude overturning as a mode of failure.

GRS–IBS consists of two abutments supporting an integrated superstructure that would function as a strut to resist overturning, and each GRS mass has a reinforced integration zone above its heel, also resisting the overturning mode of failure. Consequently, while direct sliding, bearing capacity, and global stability are evaluated in conventional ways, overturning is sometimes addressed by inspection and comparison to observations of past performance.

Observations of past performance show that the flexible, internally stabilized soil mass of GRS IBS construction, in combination with an RSF, results in more uniform stress distribution, resisting any applied vertical and lateral loads. Observations also show that, in addition to lack of overturning, the combination of vertical and lateral loads, as limited by analysis of direct sliding, bearing capacity, and global stability, does not cause excessive deformation at the face of the GRS mass or other undesirable performance.

While this combination of unique features and behaviour eliminates the need to analyse overturning as a failure mode for completed GRS–IBS, the engineer may choose to analyse for overturning during an intermediate phase of construction with consideration for the time needed for an overturning mechanism to develop and the concurrent level of loading or for project configurations different from those described herein. For example, overturning may still be a viable failure mode for abutment wing walls constructed with GRS technology if they retain soil other than reinforced soil from the abutment or opposite wing wall ' if they retain natural soil.

GRS is inherently internally stable because of the interaction between the soil and the reinforcement layers. The strength and stiffness of a GRS mass depends on the unique combination of compacted soil and reinforcement. The vertical capacity of the

GRS abutment can be determined either empirically or analytically.

Empirically, the capacity is found using a stress–strain curve specific to the combination of the reinforcement type and granular fill material. If the designer uses a combination of the materials previously tested, then the appropriate stress–strain curve can be used for design. If the designer decides to change the materials from those already tested, then a performance test can be performed to obtain an applicable stress–strain curve for the empirical method. Guidelines on how to conduct a performance test are given in appendix B.

Alternatively, the designer can predict the ultimate vertical capacity of the GRS abutment by using an analytical equation. The equation is a function of reinforcement spacing, soil strength, and soil grain size. Note that the analytical method does not predict vertical deformation. A performance test is needed to adequately predict the deformation behaviour of the GRS abutment. This design method is based on the results of many full–scale experiments and verified using case history performance data collected on several in–service GRS structures more than 20 years old.

The design of GRS–IBS is based on the following assumptions:

- The spacing of the reinforcement (12 inches or less) is a principal factor in the performance of GRS–IBS.
- A GRS mass is a composite material that is stabilized internally.
- Both the compacted granular fill and the reinforcement layers strain laterally together in response to vertical stress until the system approaches a failure condition.
- A GRS mass is not supported externally, and therefore, the facing system is not considered a structural element in design.
- Lateral earth pressure at the face of a GRS mass is not significant, eliminating connection failure as a possible limit state.
- The facing elements of a GRS mass are frictionally connected to the geosynthetic reinforcement.

- Under the prescribed granular fill and reinforcement conditions, reinforcement creep is not a concern for the sustained loads. Therefore, individual reduction factors for reinforcement creep are not necessary. Creep can be accommodated safely within the factor of safety used for design.

As described in greater detail in subsequent sections of this manual, GRS–IBS design and construction processes follow from these basic assumptions and principles.

BASIC DESIGN STEPS FOR GRS–IBS

There are nine basic steps in the design of GRS–IBS. Note that the design philosophy illustrated in this section is Allowable Stress Design (ASD). It is FHWA policy that design for all Federal–aid funded projects be conducted using the AASHTO Load and Resistance Factor Design (LRFD) methodology. Guidelines to design GRS–IBS in an LRFD format are presented in appendix C.

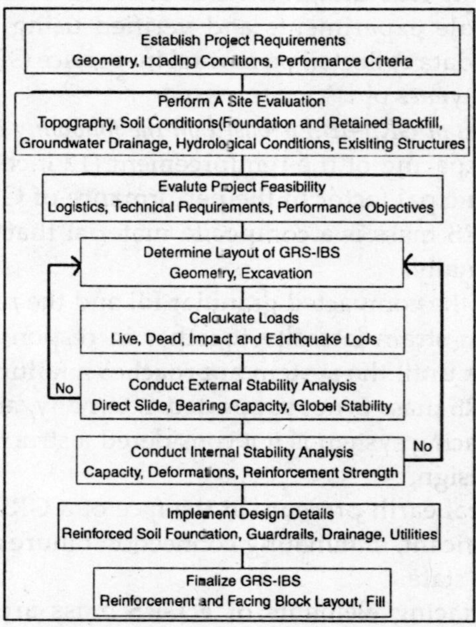

Fig. 7.1 Chart. Steps for GRS–IBS Design.

The LRFD format presented was normalized to produce the same results as the ASD method and does not represent a statistically based calibration that would be consistent with other AASHTO LRFD methods. After sufficient data is produced and collected as a result of this technology deployment and other efforts, a thorough statistical analysis will be performed to produce LRFD specifications for the design of GRS–IBS.

GRS–IBS DESIGN GUIDELINES

Step 1–Establish Project Requirements

The following parameters must be defined:

- Geometry of abutment and wing walls.
  - Height.
  - Length.
  - Batter 'vertical or near vertical.
  - Wall placement with respect to ground conditions: back slope, toe slope.
  - Skew.
  - Grade.
  - Superelevation.
- Loading conditions.
  - Soil surcharge.
  - DL.
  - LL.
  - Seismic load.
  - Impact loads.
  - Loads from adjacent structures.
- Performance criteria.
  - Design format.
  - Tolerable movements.
    - a. Vertical settlement.
    - b. Lateral displacements.
    - c. Differential settlement.
    - d. Angular distortion between abutments.
  - Design life.
  - Constraints.

a. Environmental.
b. Construction.

Step 2–Perform a Site Evaluation

To properly assess conditions at the site, a site visit must be conducted. During this visit, the following must be performed by the agency and/or its designer:

- Study the existing topography with respect to the proposed GRS–IBS.
- Check any existing structures/roads for problems to aid in the assessment and design.
- Conduct a subsurface investigation. Refer to AASHTO's *Standard Practice for Conducting Geotechnical Subsurface Investigations*. Alternatively, refer to FHWA's *Soils and Foundations Manual*.
  - Foundation soil properties (Y_f, Φ'_f, C'_f, C_u).
  - Groundwater conditions.
- Evaluate soil properties for the retained earth (Y_b, Φ_b, C'_b, C'_b).
- Evaluate soil properties for the reinforced backfill (Y_r, Φ_r, C_r, d_{max}). In addition to the basic soil properties, the maximum diameter of the granular backfill 'd_{max}' is necessary to determine the ultimate capacity and required reinforcement strength. The gradation of the reinforced backfill is also important.
- Evaluate hydraulic conditions. This can be accomplished through consultation with a qualified hydrologist.

Step 3–Evaluate Project Feasibility

The feasibility of the project should be evaluated in terms of cost, logistics, technical requirements, and performance objectives. In particular, in the case of abutments for bridges constructed over water, the potential for scour, sedimentation, and/or channel instability must be evaluated in accordance with the policy and procedures of both FHWA and AASHTO. It is necessary to determine the potential for scour at all bridges constructed over water. If the abutment will be impacted by scour, additional design requirements are necessary. These additional design requirements can be determined and implemented through a hydraulic and scour analysis of the site. Once the scour potential is determined, a countermeasure can be designed to protect the abutment against failure during a flood due to the scour that will occur at the toe of the abutment. A designed countermeasure will also protect the abutment from lateral channel migration that could undermine the foundation.

Step 4–Determine Layout of GRS–IBS

The layout of GRS–IBS is ultimately based on site conditions (desired road alignment) right of way, geotechnical issues, and hydraulic considerations. A survey should be conducted to determine the location of the GRS abutment and the layout. The layout of the abutment face wall needs to coincide with the wing walls because the system is built from the bottom up one course at a time. Both walls are built at the same time.

Use the following steps to design the abutment:

- Define the geometry of the abutment face wall and wing walls.
- Layout the abutment with respect to the superstructure skew, superelevation, grade.
  - The recommended minimum bearing width (b) for the superstructure is 2.5 ft for span lengths 'L_{span}' greater than or equal to 25 ft, as shown in figure 7.2. For span lengths (L_{span}) less than 25 ft, the minimum bearing width is 2.0 ft.
- Account for setback and clear space to calculate the elevation of the abutment face wall and the span length of the bridge.
  - The setback distance (a_b) between the back of the face and the beam seat should be the height of a standard CMU (nominally 8 inches) or more, as shown in fig. 7.2.
  - The minimum clear space (d_e), defined as the distance from the top of the uppermost facing

block to the bottom of the superstructure, should be 3 inches or 2 per cent of the abutment height, whichever is greater. The gap is to ensure that the superstructure does not bear on the facing block due to an unforeseen event.

Fig. 7.2 Bridge Seat and Setback Distances.

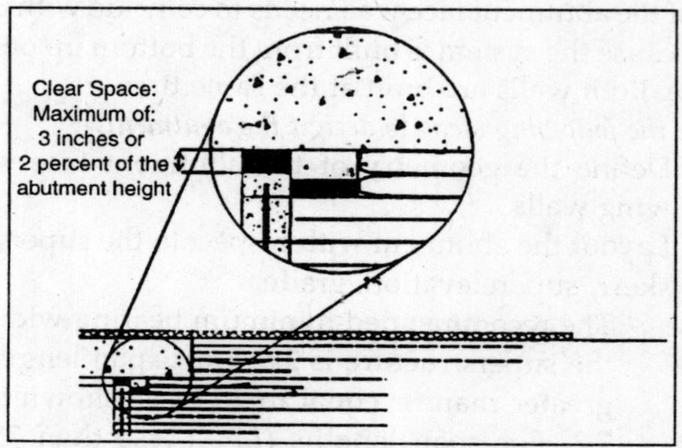

Fig. 7.3 Illustration. Clear Space Distance.

- Determine the depth and volume of excavation necessary for construction. A GRS abutment is inherently stable and therefore can be built with a truncated base to reduce the excavation. Truncation also reduces the requirements for backfill and reinforcement.
  - For span lengths (L_{span}) greater than or equal to 25 ft, a minimum base width of the wall including the block face (B_{total}) of 6 ft should initially be chosen. For span lengths (L_{span}) less than 25 ft, a

minimum base width of the wall including the block face (B_{total}) of 5 ft should initially be chosen. Whether a cut or fill situation, there should be a minimum base–to–height (B_{total}/H) ratio of 0.3. If GRS–IBS is to cross water, the base of the abutment should be placed at the calculated scour depth.

- Excavation of one–quarter the total width of the base of the abutment including the block face should be made at the base in front of the face of the wall to accommodate for construction of the RSF. The total width of the RSF should extend beyond the base of the GRS abutment by one–fourth the width of the base.
- The depth of the excavation for the RSF (D_{RSF}) should equal one–quarter the total width of the base of the GRS abutment including the block face. Additional excavation may be necessary depending on the soil conditions and should be determined by the engineer. In some situations, it may be beneficial to improve the ground beneath the RSF to reduce settlement of the bridge system.
- Before designing and constructing an RSF, it is prudent to conduct a soil investigation of the existing foundation soil including applicable lab tests to determine the soil's properties,

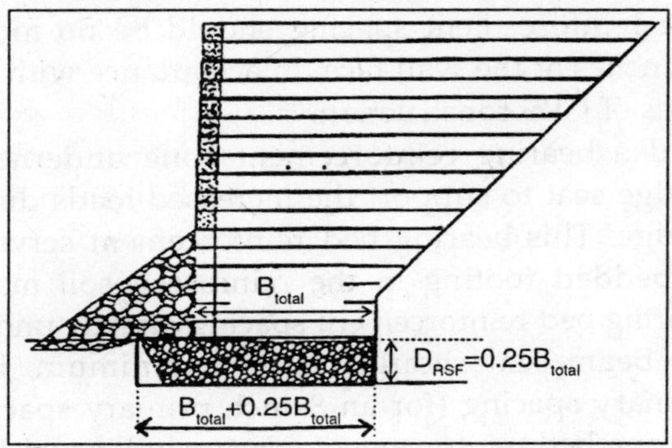

Fig. 7.4 Illustration. RSF Dimensions.

- Select the length of reinforcement for the abutment. The minimum reinforcement length at the lowest level should extend the width of the base (B_{total}) and have a minimum base–to–height ratio (*B/H*) (not including the facing block) of 0.3. The minimum reinforcement length at the lowest level should extend the width of the base (B_{total}), with a minimum of 5–6 ft or a base–to –height ratio (*B/H*) of 0.3. Once the base length of the reinforcement is chosen, the reinforcement schedule should follow the cut slope, if applicable, up to a B/H ratio of 0.7. From there, the reinforcement length can get progressively longer in reinforcement zones. Not every layer will need to extend fully to the cut slope. The progressively longer lengths of reinforcement serve to improve the quality of construction and overall stability of the GRS abutment. The reinforcement zones also serve to provide a transition from the substructure to the superstructure. The exact details of the reinforcement zones, such as number of layers and length, are left to the designer. For cut slopes flatter than 1:1, reinforcement zones with lengths larger than 1H may not be necessary. The backfill between the reinforced zone and the cut slope or retained soil must be the same structural backfill as the reinforced fill and compacted to the same effort. The reinforcement spacing should be no more than 12 inches at the wall face, in accordance with the two rules of GRS construction.
- Add a bearing reinforcement zone underneath the bridge seat to support the increased loads due to the bridge. This bearing bed reinforcement serves as an embedded footing in the reinforced soil mass. The bearing bed reinforcement spacing directly underneath the beam seat should be, at a minimum, half the primary spacing (for an 8–inch primary spacing) the bearing bed reinforcement spacing will equal 4 inches. In general, the minimum length of the bearing bed

reinforcement should be twice the setback plus the width of the bridge seat. The depth of the bearing reinforcement zone is determined based on internal stability design for required reinforcement strength. At a minimum, there should be five bearing bed reinforcement layers.

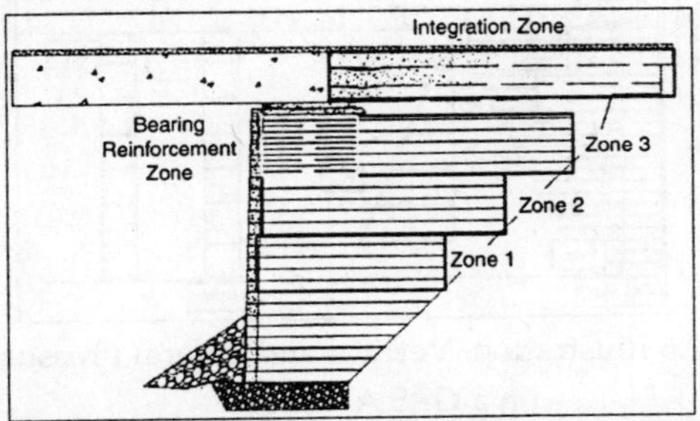

Fig. 7.5 Illustration. Reinforcement Schedule for a GRS Abutment.

- Blend the reinforcement layers in the integration zone to create a smooth transition. The layers should extend to the cut slope, if applicable, with the exception of the top reinforcement layer, depending on the site. This top layer should extend beyond the cut slope to prevent moisture infiltration. The integration zone is part of the integrated approach of GRS–IBS. It is added behind the bridge superstructure to limit the development of a tension crack at the cut slope and reinforced soil interface and to blend the approach way on to the roadway to create a smooth transition. The number of reinforcement layers in the integration zone depends on the height of the superstructure, but each wrapped layer should be no more than 12 inches in height. Additional work is needed to integrate the substructure with the superstructure within the integration zone.

Step 5–Calculate Applicable Loads

The applicable external pressures and loads (permanent and transient) on the reinforced zone of the GRS abutment should be calculated. The most common pressures (which may be resolved into forces) on GRS–IBS for stability computations are depicted in fig. 7.6.

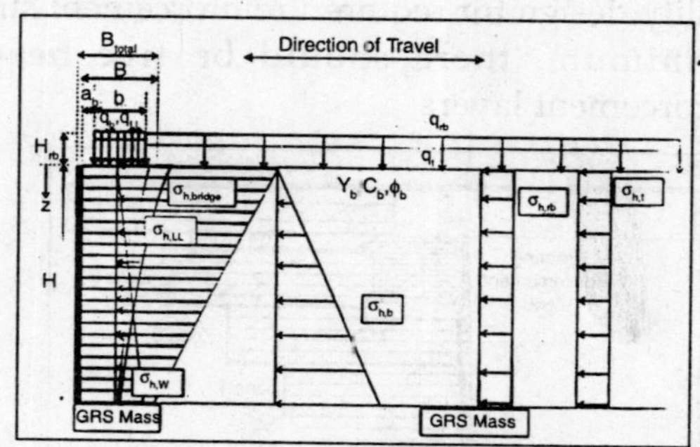

Fig. 7.6 Illustration. Vertical and Lateral Pressures on a GRS Abutment.

Table. The Applicable Pressures on a GRS Abutment are as Follows:

q_t	=	Equivalent roadway LL surcharge
$\sigma_{h,t}$	=	Lateral stress distribution due to the equivalent roadway LL surcharge
q_{rb}	=	Surcharge due to the structural backfill of the integrated approach 'road base,
$\sigma_{h, rb}$	=	Lateral stress distribution due to the structural backfill of the integrated approach
q_b	=	Equivalent superstructure DL pressure
$\sigma_{h,bridge}$	=	Llateral stress distribution due to the equivalent superstructure DL pressure
$\sigma_{h,b}$	=	Equivalent lateral stress distribution due to retained soil behind the GRS abutment
q_{LL}	=	Equivalent superstructure LL pressure
$\sigma_{h,LL}$	=	Lateral stress distribution due to the equivalent superstructure LL pressure
$\sigma_{h,W}$	=	Lateral stress distribution due to the weight of the GRS fill

LATERAL PRESSURES AND STRESSES

The lateral earth pressure can be calculated according to classical soil mechanics for active earth pressure. The active earth pressure coefficient (K_a) is calculated according to equation.

$$K_a = \frac{1-\sin\phi}{1+\sin\phi} = \tan^2\left(45° - \frac{\phi}{2}\right)$$

Where Φ is the friction angle of interest (for example) substitute Φ_b when calculating K_{ab} for the retained soil. The lateral stress distribution due to the weight of the GRS fill ($\sigma_{h,W}$) is found using Rankine's active stress condition, shown in equation.

$$\sigma_{h,w} = \gamma_r z K_{ar}$$

Where Y_r is the unit weight of the reinforced fill, z is the depth from the top of the wall, and K_{ar} is the coefficient of active earth pressure using the friction angle of the reinforced fill (Φ_r).

The lateral stress distributions due to the equivalent roadway LL surcharge ($\Sigma_{h,t}$) and structural backfill of the integrated approach ($\Sigma_{h,rb}$) are found according to equation, respectively.

$$\sigma_{h,t} = q_t K_{ab}$$

$$\sigma_{h,rb} = q_{rb} K_{ab}$$

Where q_t is the equivalent roadway LL surcharge, q_{rb} is the surcharge due to structural backfill (road base), and K_{ab} is the coefficient of active earth pressure using the friction angle of the retained backfill (Φ_b). Note that equation assume that the loading is continuous across the retained soil.

Where the loads are not continuous across the GRS abutment or retained soil, the lateral pressure is based on Boussinesq theory for load distribution through a soil mass for an area transmitting a uniform stress a distance x from the edge of the load. The actual pressure using this theory depends on the location of interest. For required reinforcement strength calculations, the location of interest is directly underneath the beam seat centerline ($x = b_q/2$ for the bridge DL).

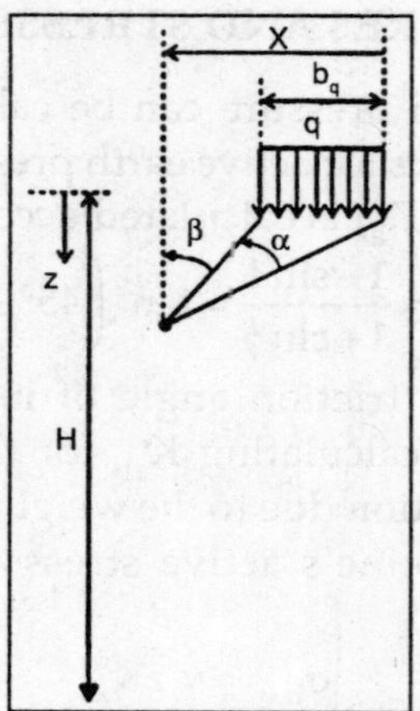

Fig. 7.7 Illustration. Boussinesq Load Distribution with Depth for a Strip Load.

The lateral pressure due to surcharge loading ($\Sigma_{h,q}$) is calculated according to equation.

$$\sigma_{h,q} = \frac{q}{\pi}[\alpha + \sin(\alpha)\cos(\alpha + 2\beta)]K_a$$

Where q is the surcharge pressure, K_a is the coefficient of activeearth pressure, and α and β are the angles shown in figure, found using equation, respectively. Note that α and β must be input in radians in equation.

$$\alpha = \tan^{-1}\left(\frac{x}{z}\right) - \beta$$

$$\beta = \tan^{-1}\left(\frac{x - b_q}{z}\right)$$

The lateral pressure in the GRS abutment due to the superstructure DL and LL will have a trend similar to that shown in figure, where the stress is highest at the top of the GRS abutment and lowest at the base. Note that the bearing bed reinforcement underneath the beam seat helps to mitigate the increased vertical (and thus lateral) pressures in this location. In fact, the bearing bed reinforcement is recommended in the design of GRS abutments for this reason.

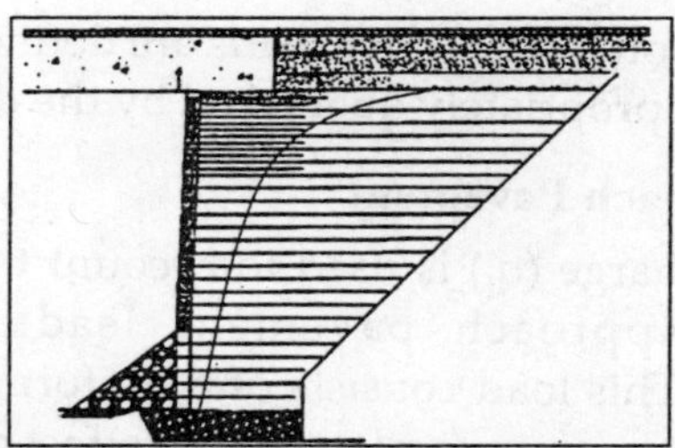

Fig. 7.8 Illustration. Internal Lateral Stress in GRS Abutment Wall Due to Bridge Loading.

Note that other load distributions are available besides Boussinesq. For example Westergaard is more applicable to a GRS mass than Boussines. However, it gives lower stresses than Boussinesq, and therefore, using Boussinesq will provide a more conservative estimate of stresses.

DEAD LOADS

Bridge

In a GRS–IBS design with adjacent concrete box beams, the bridge superstructure bears directly upon the GRS abutment. For superstructures with spread girders, a footing 'which bears directly upon the GRS abutment, is necessary to ensure even load distribution on the GRS abutment. The equivalent DL design pressure on the abutment seat includes the dead loads due to the bridge beams, asphalt, overlay, guardrail, and any other applicable permanent loads related to the superstructure.

Road Base

Behind the bridge beams, road base is wrapped in geotextile (called the integrated approach). The wrapped face controls lateral load from the road base on the beam or abutment sill.

LIVE LOADS

There are two applications of LL that affect the design of GRS–IBS: LL on the approach pavement and LL on the superstructure. Both of these live loads are defined by AASHTO and should be appropriately quantified by the design engineer.

LL on the Approach Pavement

An LL surcharge (q_t) is used to account for the vehicular load on the approach pavement leading up to the superstructure. This load consists of a uniform height (h_{eq}) of earth that produces an equivalent lateral effect on the abutment as the application of the vehicular LL specified for the superstructure. The equivalent height of earth is dependent on the abutment height and the orientation of the abutment with respect to the roadway. This load is used for both internal and external stability analyses.

LL on the Superstructure

The vehicular LL used for designing GRS–IBS is determined by applying the HL–93 LL model to the superstructure. This model consists of appropriately locating a design truck or design tandem in combination with a design lane load in each design lane of the bridge to create the maximum force effect at each abutment. The vehicular portion of the LL model is amplified for dynamic load allowance (impact). The governing LL is distributed to the abutment by multiplying by the number of design lanes and dividing by the bridge seat bearing area. This equivalent distributed LL pressure on the abutment seat (q_{LL}) can be determined using equation.

$$q_{LL} = \frac{(LL + IM)_{total}(N_{lanes})}{b(B_b)}$$

Where N_{lanes} is the number of design lanes on the bridge, b is the bridge seat bearing width, B_b is the width of the bridge, and ($LL+IM_{total}$) is the governing abutment reaction for the HL–93 LL model for one lane.

If the bridge seat bearing width is unknown and needs sizing, the LL from the superstructure should be quantified as a reaction (Q_{LL}) rather than a pressure.

$$Q_{LL} = (LL + IM)_{total}(N_{lanes})$$

DESIGN PRESSURE

Adding LL on the superstructure and bridge DL per abutment will give the total load that the bridge seat must support. Dividing this total load by the area of the bridge seat will give the bearing pressure.

For abutment applications, the bearing pressure should be targeted to around 4,000 lbs/ft^2. If this is exceeded, the width of the bridge seat should be increased. Although higher design pressures have been successfully applied to in–service GRS–IBS, this is not encouraged.

Step 6–Conduct an External Stability Analysis

The external stability of GRS–IBS is evaluated by looking at the following potential external failure mechanisms:

- Direct sliding.
- Bearing capacity.
- Global stability.

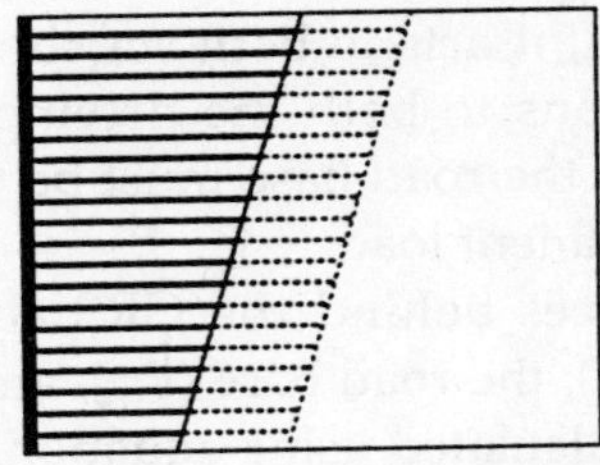

Fig. 7.9 Illustration. External Stability: Direct Sliding.

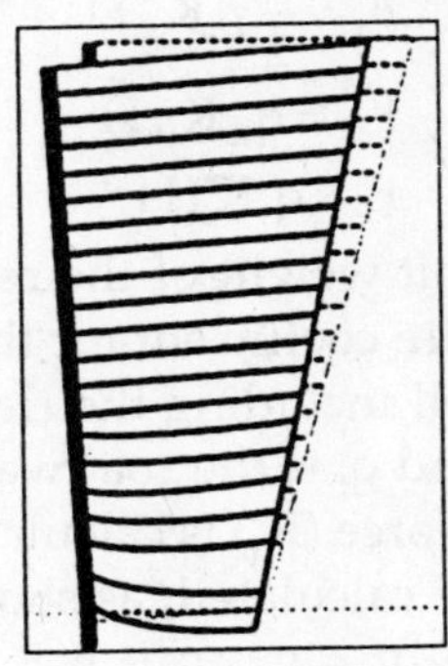

Fig. 7.10 Illustration. External Stability: Bearing Capacity.

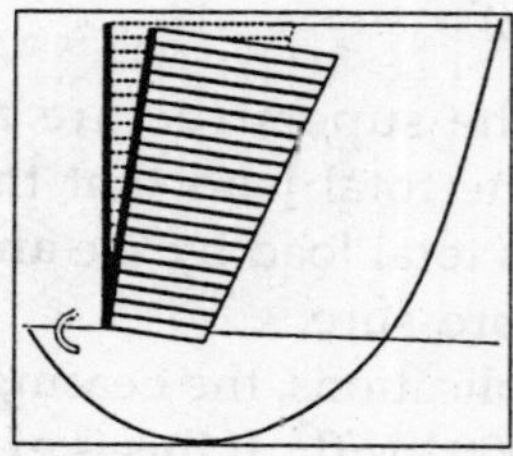

Fig. 7.11 Illustration. External Stability: Global Stability.

DIRECT SLIDING

The GRS abutment must resist translation, or direct sliding. The LL on the approach pavement (q_t) is assumed to act only over the retained backfill and not the reinforced soil mass. While the contribution of (q_t and q_{LL}) is ignored for both a wall and an abutment, the bridge load (q_b) has a stabilization effect against direct sliding when considering an abutment.

Since the road base extends over the GRS abutment and the retained backfill, it acts to both stabilize and drive direct sliding. Contributions to both the driving force and to the resisting force from the road base must be taken into account because it is a permanent load.

The thrust forces behind the GRS abutment from the retained backfill (F_b), the road base (F_{rb}), and the roadway LL surcharge (F_t) are calculated using equation.

$$F_b = \frac{1}{2}\gamma_b K_{ab} H^2$$

$$F_{rb} = q_{rb} K_{ab} H$$

$$F_t = q_t K_{ab} H$$

Where Y_b is the unit weight of the retained backfill, K_{ab} is the active earth pressure coefficient for the retained backfill, H is the height of the wall including the clear space distance, q_{rb} is the road base DL, and q_t is the roadway LL.

The total driving force (F_n) is calculated by summing each thrust force previously calculated, as shown in equation.

$$F_n = F_b + F_{rb} + F_t$$

The resisting force (R_n) is calculated according to equation.

$$R_n = W_t \mu$$

Where W_t is the total resisting weight, μ is the friction factor between the wall base and the foundation (taken as tan Φ_{crit}), and Φ_{crit} is the critical friction angle. Since the RSF is encapsulated with geotextile, sliding at the base of the GRS abutment will occur between soil and the geotextile reinforcement. The critical friction angle will therefore be the interface friction angle between the soil and reinforcement. The interface friction angle should be determined with an interface direct shear test for the particular combination of geosynthetic and reinforced fill material (ASTM D5321).

If this information is not available for geotextiles and geogrids, assume that the friction factor is equal to $^2/_3$ times the tangent of the reinforced granularfill friction angle '$\mu = {}^2/_3 \tan' \Phi_r$.

$$W_t = W + q_b b + q_{rb} b_{rb,t}$$

Where W is the weight of the GRS abutment, q_b is the bridge DL, b is the width of the bridge load (measured along the direction of the roadway), q_{rb} is the road base DL, and $b_{rb,t}$ is the width over the GRS abutment where the road base DL acts. The LL on the approach pavement and the superstructure are not included as resisting forces because they are transient loads.

$$W = \gamma_r HB$$

Where Y_r is the unit weight of the reinforced fill, H is the height of the GRS abutment including the clear space distance, and Bis the base width of the GRS abutment not including the wall facing. The factor of safety against direct sliding (FS_{slide}) is computed according to equation. The factor of safety must be greater than or equal to 1.5. If not, consider lengthening the reinforcement at the base.

$$FS_{slide} = \frac{R_n}{F_n} \geq 1.5$$

BEARING CAPACITY

To prevent bearing failure, the vertical pressure at the base of the RSF must not exceed the allowable bearing capacity of the underlying soil foundation. The vertical pressure is a result of the weight of the GRS abutment, the weight of the RSF, the bridge seat load, the LL on the superstructure, and the LL on the approach pavement. The pressure at the base ($\sigma_{v,base,n}$) is calculated according to a Meyerhof–type distribution, shown in equation.

$$\sigma_{v,base,n} = \frac{\sum V}{B_{RSF} - 2e_{B,n}}$$

Where ΣV is the total vertical load on the GRS abutment, B_{RSF} is the width of the RSF, and $e_{B,n}$ is the eccentricity of the resulting force at the base of the wall.

$$\sum V = W + W_{RSF} + W_{face} + b_{rb,t}(q_t + q_{rb}) + b(q_b + q_{LL})$$

Where W is the weight of the GRS abutment, W_{RSF} is the weight of the RSF, W_{face} is the weight of the facing elements, $q_t$is the roadway LL, $b_{rb,t}$ is the width of the traffic and road base load over the GRS abutment, q_{rb} is the road base surcharge, q_b is the bridge DL, b is the width of the bridge seat, and q_{LL} is the LL on the superstructure.

$$e_{B,n} = \frac{\sum M_D - \sum M_R}{\sum V}$$

Where ΣM_D is the total driving moment, ΣM_R is the total resisting moment, and ΣV is the total vertical load. The moments should be calculated about the bottom and center of the RSF for the specific layout of the GRS abutment. If $e_{B,n}$ is negative, take $e_{B,n}$ equal to zero for the term B_{RSF}–$2e_{B,n}$.

The bearing capacity of the foundation (q_n) can be found using equation.

$$q_n = c_f N_c + \frac{1}{2} B' \gamma_f N_\gamma + \gamma_f D_f N_q$$

Where c_f is the coehsion of the foundation soil, N_c, N_y, and N_q are dimensionless bearing capacity coefficients as shown in table, Y_f is the unit weight of the foundation soil, B′ is the effective foundation width (equal to B_{RSF}–$2e_{B,n}$), and D_f is the depth of embedment. The friction angle in table should be taken as the foundation's friction angle (φ_f). If groundwater is present, modifications to equation may be necessary and are provided by AASHTO.

Table. Bearing Capacity Factors

φ_f	N_c	N_q	N_y	φ_f	N_c	N_q	N_y
0	5.14	1.0	0.0	23	18.1	8.7	8.2
1	5.4	1.1	0.1	24	19.3	9.6	9.4
2	5.6	1.2	0.2	25	20.7	10.7	10.9
3	5.9	1.3	0.2	26	22.3	11.9	12.5
4	6.2	1.4	0.3	27	23.9	13.2	14.5
5	6.5	1.6	0.5	28	25.8	14.7	16.7
6	6.8	1.7	0.6	29	27.9	16.4	19.3
7	7.2	1.9	0.7	30	30.1	18.4	22.4
8	7.5	2.1	0.9	31	32.7	20.6	26.0
9	7.9	2.3	1.0	32	35.5	23.2	30.2
10	8.4	2.5	1.2	33	38.6	26.1	35.2
11	8.8	2.7	1.4	34	42.2	29.4	41.1
12	9.3	3.0	1.7	35	46.1	33.3	48.0
13	9.8	3.3	2.0	36	50.6	37.8	56.3
14	10.4	3.6	2.3	37	55.6	42.9	66.2
15	11.0	3.9	2.7	38	61.4	48.9	78.0
16	11.6	4.3	3.1	39	67.9	56.0	92.3
17	12.3	4.8	3.5	40	75.3	64.2	109.4
18	13.1	5.3	4.1	41	83.9	73.9	130.2
19	13.9	5.8	4.7	42	93.7	85.4	155.6
20	14.8	6.4	5.4	43	105.1	99.0	186.5
21	15.8	7.1	6.2	44	118.4	115.3	224.6
22	16.9	7.8	7.1	45	133.9	134.9	271.8

The factor of safety against bearing failure ($FS_{bearing}$) is computed according to equation. The factor of safety must be greater than or equal to 2.5.

If not, increase the width of the GRS abutment and RSF (by increasing the length of the reinforcements), replace the foundation soil with a more competent soil, or add embedment depth.

$$FS_{bearing} = \frac{q_n}{\sigma_{V,base,n}} \geq 2.5$$

Beyond bearing capacity, consolidation settlement should be evaluated to ensure excessive deformations will not occur over the life of the bridge. Design considerations such as excavation and the RSF reduce the pressure on the foundation soil.

Nevertheless, settlement of the foundation soil should be assessed as with any other spread footing according to FHWA guidance Determining the criterion for tolerable foundation settlement is left up to the engineer. A stress history analysis should be conducted to ascertain settlement and stability prediction.

Answers to the following questions will provide insight on the stress history for an efficient design:

- Is the site bridge a replacement project built in the same location?
- What was the performance of the existing bridge?
- Were there any chronic maintenance issues associated with the existing structure?
- What was the combined weight of the abutment and superstructure within the footprint of the new bridge foundation? How does that stress compare with the stress of the new structure?
- Does the site involve an excavation equivalent to the weight of the new GRS–IBS?
- Can the new bridge be built behind the existing foundation?

GLOBAL STABILITY

Global stability is evaluated according to classical slope stability theory using either rotational or wedge analysis. To facilitate the global stability check, it is prudent to collect accurate soil property information. Standard slope stability computer programmes can then be used to assess the global and compound stability of a GRS structure. The factor of safety for global stability should equal at least 1.5.

Step 7–Conduct Internal Stability Analysis

The internal stability analysis will vary slightly depending on the whether ASD or LRFD is the chosen design method. ASD is presented. For guidance on LRFD, refer to appendix B.

ULTIMATE CAPACITY

The ultimate vertical capacity of a GRS abutment is found either empirically or analytically. It is recommended that the ultimate capacity be found empirically if possible. A performance test should be conducted to determine the ultimate capacity if the reinforced fill is different from those used in the performance tests reported in this guide. Testing will provide the most accurate results for the design. If a performance test cannot be performed, the analytical method can be used to determine the ultimate capacity.

Empirical Method

Empirically, the results of an applicable performance test using the same geosynthetic reinforcement and compacted granular backfill as planned for the site should be used. The *ultimate vertical capacity* in this case is defined as the stress at which the performance test mass strains 5 per cent vertically. The ultimate vertical capacity is found in figure 7.12. For this performance test, the nominal capacity ($q_{ult,emp}$) is equal to 26 ksf for a vertical strain of 5 per cent.

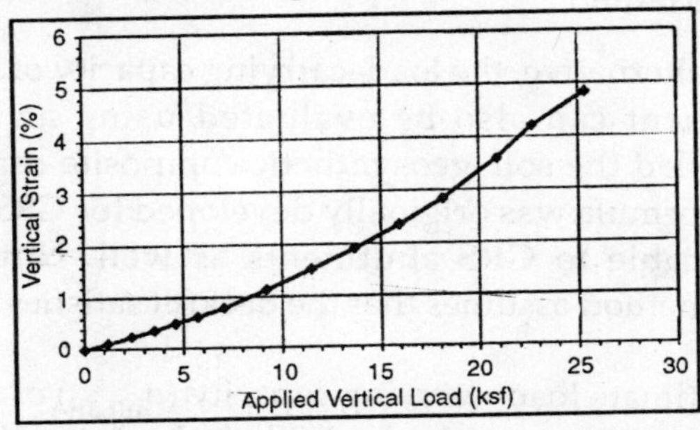

Fig. 7.12 Graph. Design Envelope for Vertical Capacity and Strain at 8–Inch Reinforcement Spacing.

Note that fig. 7.12 represents the load–settlement performance of a GRS structure with reinforcement spaced at 8 inches, well–compacted AASHTO No. 89 fill material 'having a friction angle of 48 degrees and no cohesion,, and 4,800 lb/ft woven PP geosynthetic reinforcement. Other materials have also been tested and are shown in the synthesis report.

If the materials used are outside the recommendations provided, then a performance test must be performed to obtain the applicable stress–strain curve similar to figure. The total allowable pressure on the GRS abutment ($V_{allow,\ emp}$) is the ultimate capacity ($q_{ult,emp}$) divided by a factor of safety for capacity ($FS_{capacity}$) of 3.5, as shown in equation.

$$V_{allow,emp} = \frac{q_{ult,emp}}{FS_{capacity}} = \frac{q_{ult,emp}}{3.5}, 23,$$

The applied vertical stress '$V_{applied}$'' which is equal to the unfactored sum of the vertical pressures on the bridge bearing area, must be less than $V_{allow,emp}$. This includes the DL from the bridge 'q_b, and the LL on the superstructure 'q_{LL}. The DL due to the road base 'q_{rb}, and the LL due to the approach pavement 'q_t, are located behind the bearing area and are therefore not included in vertical capacity calculations related to the bridge superstructure.

$$V_{applied} = q_b + q_{LL} \leq V_{allow,emp}$$

Analytical Method

As an alternative, the load–carrying capacity of a GRS wall and abutment can also be evaluated using an analytical formula called the soil–geosynthetic composite capacity. The analytical formula was originally developed for GRS walls, but it is applicable to GRS abutments as well. Note that the analytical method assumes that the backfill satisfies the criteria outlined.

The ultimate load–carrying capacity ($q_{ult,an}$) of a GRS wall constructed with a granular backfill can be determined by the soil–geosynthetic composite capacity equation shown in equation.

$$q_{ult,an} = \left[0.7^{\left(\frac{S_V}{6d_{max}}\right)} \frac{T_f}{S_V} \right] K_{pr}$$

Where S_v is the reinforcement spacing, d_{max} is the maximum grain size of the reinforced backfill, T_f is the ultimate strength of the reinforcement, and K_{pr} is the coefficient of passive earth pressure for the reinforced fill.

$$K_{pr} = \frac{1+\sin\phi r}{1-\sin\phi r} = \tan^2\left(45° + \frac{\phi_r}{2}\right)$$

Where Φ_r is the friction angle of the reinforced backfill. The friction angle should be determined from a large–scale direct shear device (ASTM D3080).

The total allowable pressure on the GRS abutment ($V_{allow,an}$) is the ultimate capacity found analytically ($q_{ult,an}$) divided by a factor of safety for capacity '$FS_{capacity}$', of 3.5.

$$V_{allow,an} = \frac{q_{ult,an}}{FS_{capacity}} = \frac{q_{ult,an}}{3.5}$$

The applied vertical stress ($V_{applied}$), which is equal to the unfactored sum of the vertical pressures on the bridge bearing area, must be less than $V_{allow,an}$. This includes the DL from the bridge (q_b) and the equivalent LL on the bridge (q_{LL}). The DL due to the road base (q_{rb}) and the LL due to the approach pavement (q_t) are located behind the bearing area and are therefore not included in capacity calculations related to the bridge superstructure.

$$V_{apllied} = q_b + q_{LL} \leq V_{allow,an}$$

DEFORMATIONS

The approach for determining vertical deformation involves empirically finding the strain from an applicable performance test curve. If the materials used are within the specifications, then the curve shown in fig. 7.12 can be used. Otherwise, a performance test must be conducted. The lateral strain is then determined analytically assuming the theory of zero volume change.

Vertical

The vertical strain of the GRS abutment is found from the intersection of the applied vertical stress due to the DL (q_b) and the performance test design envelope for vertical strain. The vertical strain should be limited to 0.5 per cent unless the engineer decides to permit additional deformation. The vertical deformation, or settlement, of the GRS abutment is the vertical strain multiplied by the height of the wall or abutment. Because the GRS abutment is built with a granular fill, the majority of settlement within the GRS abutment will occur immediately after the placement of DL (q_b) and before the bridge is opened to traffic.

The settlement of the underlying foundation soils is determined separately using classic soil mechanics theory for immediate (elastic) and consolidation settlement. Factors such as excavation and the RSF should be taken into account, as the removal of overburden relieves stress on the foundation soil. Settlement of the foundation soil can be calculated using the FHWA *Soils and Foundations Reference Manual.*

Lateral

In response to a vertical load, the composite behaviour of a properly constructed GRS mass is such that both the reinforcement and soil strain laterally together. This fact can be used to predict both the maximum lateral reinforcement strain and the maximum face deformation at a given load. The method conservatively assumes a zero volume change in the GRS abutment, which represents a worst–case scenario. The maximum lateral displacement of the abutment face wall can be estimated using equation. The lateral strain (ε_L) is then found using equation should be limited to 1 per cent.

$$D_L = \frac{2b_{q,vol}D_v}{H}$$

$$\varepsilon_L = \frac{D_L}{b_{q,vol}} = \frac{2D_v}{H} = 2\varepsilon_v$$

Where $b_{q,vol}$ is the width of the load along the top of the wall (including the setback), D_v is the verticalsettlement in the GRS abutment, H is the wall height including the clear space distance, and ε_V is the vertical strain at the top of the wall. Note that equation come from the assumptions of a triangular lateral deformation and a uniform vertical deformation. This assumption is based on observed deformation behaviour of GRS. Also note that the location of the maximum lateral deformation depends on the loading and fill conditions, but the volume gained will still equal the volume lost. The maximum deformation of a GRS abutment often occurs in the top third of the abutment/wall.

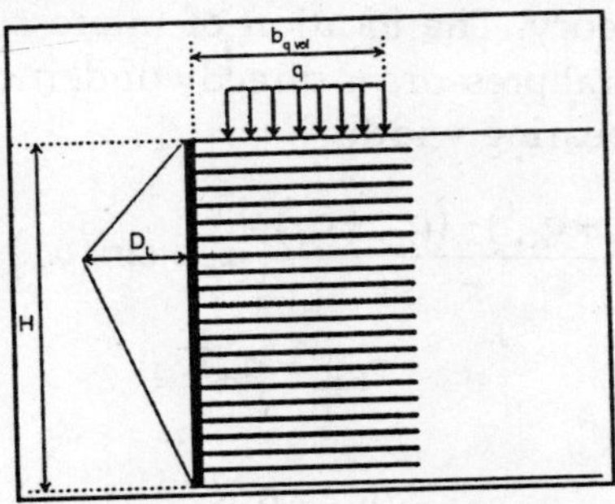

Fig. 7.13 Illustration. Lateral deformation of a GRS structure.

REQUIRED REINFORCEMENT STRENGTH

The required reinforcement strength in the direction perpendicular to the wall face (T_{req}) can be determined analytically by equation. The required reinforcement strength should be calculated at each layer of reinforcement to ensure adequate strength throughout the GRS abutment.

$$T_{req} = \left[\frac{\sigma_h}{0.7\left(\frac{S_v}{6d_{max}}\right)} \right] S_v$$

Where S_V is the reinforcement spacing, d_{max} is the maximum grain size of backfill, and σ_h is the total lateral stress within the GRS abutment at a given depth and location.

$$\sigma_h = \sigma_{h,W} + \sigma_{h,bridge,eq} + \sigma_{h,rb} + \sigma_{h,t}$$

Where $\sigma_{h,W}$ is the lateral earth pressure using Rankine's active stress condition, $\sigma_{h,bridge,eq}$ is the lateral pressure due to the equivalent bridge load, $ó_{h,rb}$ is the lateral pressure due to the road base, and $\sigma_{h,t}$ is the lateral pressure due to the roadway LL. To simplify calculations, the approach LL and road base DL are extended across the abutment. The vertical components of these loads are then subtracted from the bridge DL and LL, giving an equivalent bridge load. The lateral stress due to the equivalent bridge load is then calculated according to Boussinesq theory. The location of interest to determine the maximum lateral pressure is directly underneath the centerline of the bridge bearing width.

$$\sigma_{h,bridge,eq} = \frac{(q_b + q_{LL}) - (q_{rb} + q_t)}{\pi}\left[\alpha_b + \sin(\alpha_b)\cos(\alpha_b + 2\beta_b)\right]K_{ar}$$

$$\sigma_{h,rb} = q_{rb}K_{ar}$$

$$\sigma_{h,t} = q_t K_{ar}$$

Where q_b, q_{rb}, q_t, and q_{LL} are the bridge DL, road base DL, roadway LL, and bridge LL surcharges,respectively, and α_b and β_b are the angles shown in figure, found using equation respectively.

$$\alpha_b = \tan^{-1}\left(\frac{b}{2z}\right) - \beta b$$

$$\beta b = \tan^{-1}\left(\frac{-b}{2z}\right)$$

The required reinforcement strength 'T_{req}, must satisfy two criteria: (1) it must be less than the allowable reinforcement strength (T_{allow}), and (2) it must be less than the strength at 2 per cent reinforcement strain ($T\varepsilon$=2%)

In design, a minimum value of the ultimate reinforcement strength (T_{allow}) is needed to ensure adequate ductility and satisfactory long–term performance. In addition, it is prudent to specify the resistance required at the working load, ($T@\varepsilon$=2%)to ensure satisfactory performance under the in–service condition.

For abutments, a minimum ultimate tensile strength (T_f) of 4,800 lb/ft is required. The allowablereinforcement strength (T_{allow}) is found by applying a factor of safety for reinforcement strength (FS_{reinf}) of 3.5 to the ultimate strength. The required reinforcement strength (T_{req}) must be less than T_{allow}.

$$T_{allow} = \frac{T_f}{FS_{reinf}} = \frac{T_f}{3.5}$$

Since geosynthetic reinforcements of similar strength can have rather different load–deformation relationships depending on the manufacturing process and the polymer used, it is important that T_{req} be less than the strength at 2 per cent reinforcement strain. The strength of the reinforcement at 2 per cent, ($T_{@\varepsilon} = 2\%$) is often given by the geosynthetic manufacturer. If T_{req} is greater than $T_{@\varepsilon} = 2\%$, a different geosynthetic must be chosen, the ultimate strength must be increased, or the reinforcement spacing must be decreased.

While the strength of the reinforcement can theoretically vary along the height of the GRS abutment, it is recommended that only one strength of reinforcement be used throughout the entire abutment. This simplifies the construction process and avoids placement errors for the reinforcement.

Depth of Bearing Bed Reinforcement

The required reinforcement strength (T_{req}) is found at each 8–inch primary spacing layer. If T_{req} is greater than the allowable reinforcement strength (T_{allow}) or the strength at 2 per cent strain ($T_{@\varepsilon} = 2\%$), then the reinforcement spacing must be reduced to 4 inches to the depth at which T_{req} is less than T_{allow} or $T_{@\varepsilon}$=2%.This depth is termed the *bearing reinforcement bed*. The minimum required depth is five courses of block.

To check that 4–inch spacing for the bearing reinforcement bed is adequate, calculate the required reinforcement strength again for this new spacing in the top layers to ensure that T_{req}

is less than T_{allow} and $T_{@\varepsilon} = 2\%$ for all layers throughout the GRS abutment.

Step 8–Implement Design Details

Fig. 7.14 are typical cross sections of a GRS wall (or wing wall) and an abutment face wall and illustrate design details that will be discussed in this section.

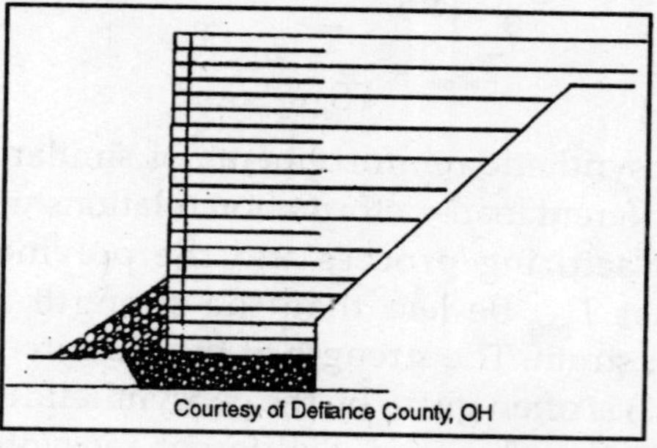

Fig. 7.14 Illustration. Typical Cross Section of a GRS Wing Wall.

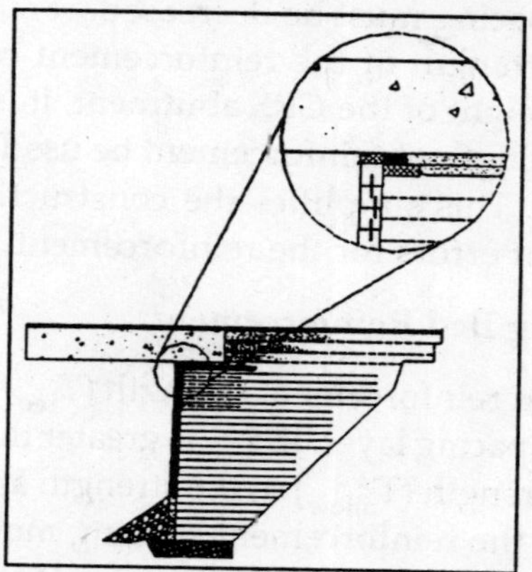

Fig. 7.15 Illustration. Typical Cross Section of a GRS Abutment Face Wall.

In the case of an abutment, finalize the design layout for ease of construction, drainage, and other considerations that might affect the performance, serviceability, or efficiency of design.

The following are some GRS design implications and related details for consideration:

- Conduct a hydraulic analysis in accordance with all appropriate regulatory and policy guidance. Consult a licensed hydraulic engineer if necessary.
- Ensure that the face of the abutment (which includes the parapet) is wide enough to accommodate the installation of guardrails. The additional width should be enough to allow the guardrail to lay down. This lay–down length is approximately 4 ft. Steel rail posts should be used because wooden posts are nearly impossible to drive into the GRS mass.

Fig. 7.16 Guardrail lay–down Distance.

- Consider a core of native soil in the center of the abutment face and two adjacent wing walls to minimize excavation. The wing walls can be truncated like the abutment. Extend the wing walls sufficiently into the cut slope to prevent erosion caused by undermining or piping. This should be a minimum of two facing–block lengths.

Fig. 7.17 GRS Abutment and Wing Walls Built Around Core of Native Soil

- Determine whether to build wing walls with either a full face or a stepped face that leads into the cut slope. The decision depends on several factors related to the height of the abutment, grade of fill slope, and time and materials. For abutments less than 12 ft in height, a full face is probably most efficient as it is the easiest to construct. However, for abutments greater than 12 ft, it might be more efficient to design a stepped –face wall that leads into the cut slope. Stepped walls use less material but require additional labour in building the second foundation to support the extended stepped wall. In either case, all facing blocks should be supported on well–compacted structural fill.
- Include channel drains along the wings walls to facilitate run-off. The drain path should not be located directly against the wall face. Armor the drain path with a strip of geotextile beneath a layer of channel rock. Grade in compacted native soil against the wing walls with a slope leading to the drainage path.
- GRS–IBS has been used for bridges with skew, superelevation, and grade without problems or serviceability issues.
  - For a skewed bridge, it is important to maintain the minimum bearing area of 2.5 ft along the length of the abutment face wall.
  - For a bridge with superelevation, it is important to ensure that the minimum number of bearing bed reinforcement layers beneath the beam seat (calculated in step 7) are installed across the length of the abutment face.
  - At this time, there are no special considerations for opposing GRS abutments that support a bridge on a grade.
- Contain the GRS integrated approach fill by wrapping the geotextile layers adjacent to the beam ends to prevent lateral spreading Extend the reinforcement layers at the approach back onto the road, as indicated in fig. 7.17.

- Avoid any abrupt transition of soil type from the roadway to the bridge. While the RSF and abutment should use a reinforcement with a minimum ultimate strength of 4,800 lb/ft, a lighter geosynthetic of about 2,400 lb/ft could be used for the integrated approach. However, it is recommended that only one strength of reinforcement be used on site to simplify the construction process and avoid placement errors.
- Plan ahead to avoid trenching, and account for the possible installation of utilities.
- Locate and plan to accommodate existing and potential future utilities.

Step 9–Finalize Material Quantities and Layout

To develop the reinforcement schedule, choose a reinforcement length that makes use of the entire roll of the reinforcement material. Reinforcement material is usually 12– to 18–ft wide. For example, the width of a PP geosynthetic roll is 12 ft, and the base of a GRS wall is 6 ft including the width of the wall face. The roll can be cut in half by a chainsaw, and a 6–ft–wide roll can be used to build the base of the wall. The remaining 6–ft–wide rolls can be used for secondary or intermediate layers of reinforcement in the walls.

Draw the layout to scale to avoid errors in the calculation of quantities. Add 10 per cent to the estimate of all materials. When using CMU, use the exact dimensions of $7^5/_8$ inches by $7^5/_8$ inches by $15^5/_8$ inches and buy both corner and face blocks.

Building GRS abutments vertically without a batter eliminates the need to trim blocks. This will make it more difficult to hide lateral movement and may give an illusion of instability when the structure is, in fact, stable. Use only high–quality, well–graded gravel.

BOWMAN ROAD BRIDGE, DEFIANCE COUNTY, OH

Construction of Bowman Road Bridge was completed in October 2005 by a Defiance County, OH,construction crew. This project represents the initial deployment of GRS–IBS. The structure was chosen for a design example because it demonstrates many of the variables that can be accommodated by GRS–IBS technology and illustrates the versatility of the construction method.

ESTABLISH PROJECT REQUIREMENTS

GRS–IBS was used for the Bowman Road Bridge project. The project included an abutment and a wing wall on each side of the bridge. A top view of the proposed project is shown in fig. 7.18 is an aerial view of the site with the proposed bridge superimposed.

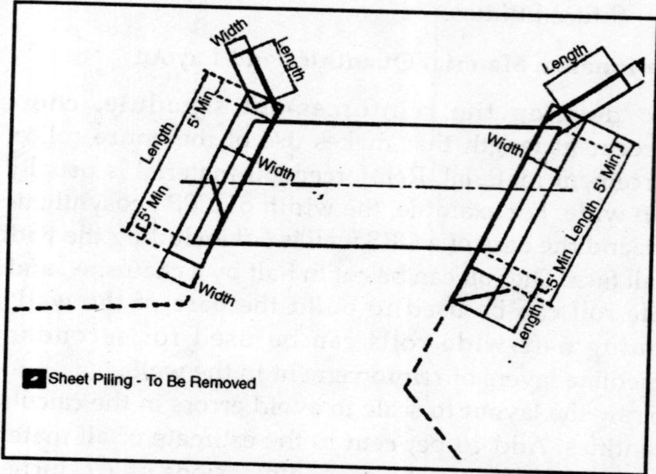

Fig. 7.18 Illustration. Top View of Bowman Road Bridge Showing the Bridge, Abutments, and Wing Walls.

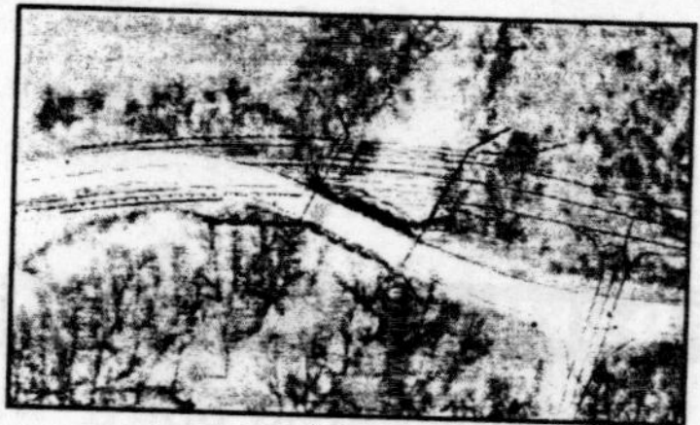

Fig. 7.19 Illustration. Aerial view of the Existing site with the Planned Bowman Road Bridge Superimposed.

Schematics of the proposed abutments are shown in fig. 7.19.

The project requirements are as follows:

- Geometry.
  - *Wall height (H_{abut})*: 15.25 ft.
  - *Abutment length (L_{abut})*: 43.6 ft.
  - *Bridge width (B_b)*: 34 ft.
  - *Batter (ω)*: 2 degrees.
  - *Wall placement with respect to ground conditions (back slope, toe slope)*: None.
  - *Skew (S_k)*: 24 degrees.
  - *Grade (G)*: 0.006 ft/ft.
  - *Superelevation (S_e)*: 7.6 degrees.
- Loading Conditions.
  - *Soil surcharge*: Road base will be placed behind the bridge beam to create a smooth transition.
  - *DL*: DL includes the weight of the bridge beam along with any corresponding components.
  - *LL*: LL includes traffic and truck loads which are simulated as a surcharge.
  - *Seismic load*: Seismic effects are negligible in this area.
  - *Impact loads*: No impact loads are considered.
  - Loads from adjacent structures: Not applicable.
- Performance Criteria.
  - *Design code*: ASD.
  - Tolerable movements.
    a. Vertical settlement: Vertical strain (ε_V) is limited to 0.5 per cent.
    b. Lateral displacements: Lateral strain (ε_L) is limited to 1 per cent.
  - *Design life*: 100 years.
  - Constraints.
    a. *Environmental*: None.
    b. *Construction*: Sheet piling from the existing bridge remains in place, reducing the need for

two wing walls in the IBS. Only one wing wall is required.

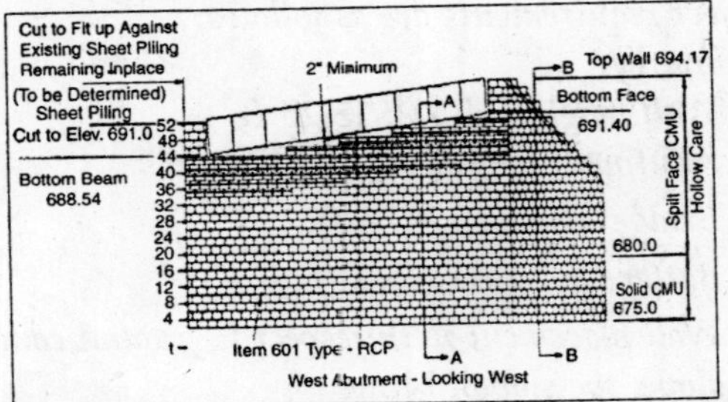

Fig. 7.20 Illustration. Schematic of the West Abutment for Bowman Road Bridge.

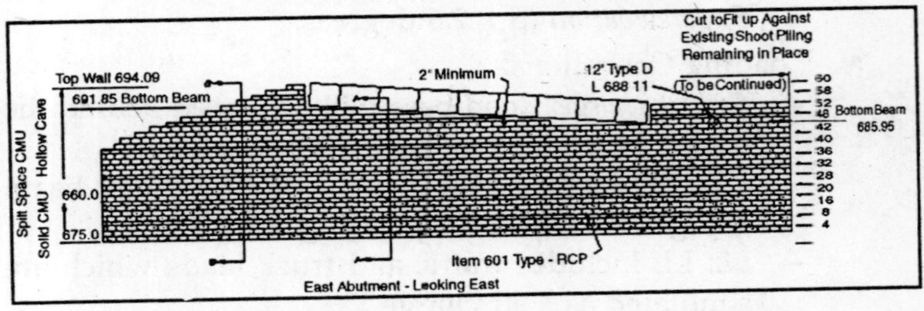

Fig. 7.21 Illustration. Schematic of the East Abutment for Bowman Road Bridge.

SITE EVALUATION

The previous bridge at the site was replaced because it was functionally obsolete and structurally deficient. The previous bridge did not experience any problems related to settlement or excessive deformations due to the site conditions. However, a sheet pile wall was installed to protect the stone wall abutments from erosion. The site evaluation determined that the existing sheet piling should remain in place to support the stream bank. This eliminated the need for wing walls on one side adjacent to the old bridge.

The replacement structure required realignment to meet current road design standards for roadway safety because the location had been prone to accidents. The new Bowman Road

Bridge crosses Powell Creek. The proposed location of the new abutments adjacent to the old bridge was not expected to cause any problems with the stream flow.A hydraulic analysis confirmed that the existing bridge did not have any appreciable potential scour. Therefore, an RSF with appropriate scour countermeasures (in this case, riprap) was used.

A subsurface evaluation was conducted by performing standard penetration tests (SPTs) near the site. The physical characteristics of the soil were determined through index tests taken on split spoon samples.

The foundation soil at the site was an overconsolidated clay (with intermediate layers of sandy silt and gravels) with N–values greater than 50 blows per ft at the elevation of the bottom of the abutment.

Local experts indicated the clay had historically been preloaded with a nearly 1–mi–thick sheet of ice. The clay in this region is also known to be fat and sticky when wet. The bearing capacity of the stiff clay had not been a problem in past projects in the area.

The N–value of the foundation soil can be correlated into an undrained shear strength using published guidance. For blow counts greater than 30 blows per ft, the unconfined compressive strength is greater than 8,000 lb/ft^2. The undrained shear strength is therefore estimated as at least 4,000 lb/ft^2. The design properties for the foundation soil are shown in table.The retained backfill is composed of the same material as the foundation soil.

Table. Foundation and Retained Backfill Soil Properties.

Property	Notation	Measurement
Foundation and backfill soil unit weight	γ_f, γ_b	120 lb/ft^3
Foundation and backfill soil undrained shear strength	c_u, c_b	4,000 lb/ft^2
Foundation and backfill soil effective cohesion	c'_f, c'_b	400 lb/ft^2
Foundation and backfill soil effective friction angle	Φ'_f, Φ'_b	28 degrees

The road base was a granular fill material that was brought to the site. For the Bowman Road Bridge project, the properties of the road base are given in table.

Table. Road base Soil Properties.

Property	Notation	Measurement
Road base unit weight	Y_{rb}	140 lb/ft^3
Road base cohesion	c_{rb}	0 lb/ft^2
Road base friction angle	Φ_{rb}	40 degrees

The reinforced fill for the GRS abutment was a select granular fill AASHTO No. 89 stone. Testing was performed on this fill to determine the *c* and Φproperties. The properties of this fill are provided in table.

Table. Reinforced Fill Properties.

Property	Notation	Measurement
Reinforced fill unit weight	Y_r	110 lb/ft^3
Maximum diameter of rei -nforced fill	d_{max}	0.5 inches
Reinforced fill cohesion	c_r	0 lb/ft^2
Reinforced fill friction angle	Φ_r	48 degrees

EVALUATE PROJECT FEASIBILITY

As mentioned in step 2, scour was not a significant concern for this bridge. The project was therefore considered feasible for this site. Scour protection was added as a precaution. The riprap was sized for 8.8–10.2 ft/s to create a scour protection apron adjacent to and in front of the abutment face and wing walls. Prior to placement, a 5– to 8–ft–wide strip of geotextile reinforcement between the face of the RSF and the riprap was pinned under the first course of facing blocks to secure it in place. The purpose of the geotextile reinforcement was to create a barrier to mitigate loss of soil beneath the riprap.

DETERMINE LAYOUT OF GRS–IBS

- Define the geometry of the abutment face wall and wing walls:

- Layout the abutment with respect to the superstructure:. The distance between the abutment faces was 72 ft. Therefore, since the length of the bridge was greater than 25 ft, the minimum bearing width 'b, for the superstructure was 2.5 ft. A bearing width of 4 ft, however, had been chosen for this bridge.
- Account for setback and clear space: The bridge seat had a setback of 8 inches from the edge of the wall. The clear space was 4 inches, which was greater than 2 per cent of the wall height.
- Determine the depth and volume of excavation necessary for construction:
  - A base width of the wall including the block face of 6 ft was chosen for this abutment since the span length was greater than 25 ft and 0.3H was less than the 6–ft minimum. Subtracting the wall face width '7.625 inches,, the reinforcement length at the base of the wall was 5.4 ft. This equates to a B/H ratio of 0.35, which is greater than the minimum B/H ratio of 0.3.
  - Excavation of 1.5 ft 'one–quarter the width, including the block face, was made at the base in front of the face of the wall to accommodate for construction of the RSF. The total width of the RSF was therefore 7.5 ft.
  - The depth of the excavation for the RSF was equal to one–quarter the width of the base including the block face–1.5 ft.
- Select the length of reinforcement for the abutment: The reinforcement length at the base of the wall was equal to 6 ft or 5.4 ft not including the reinforcement necessary for the frictional connection. The reinforcement lengths up the wall were chosen based on the cut slope angle and an optimization of the width of the reinforcement rolls. The reinforcement schedule is shown in figure 7.19.

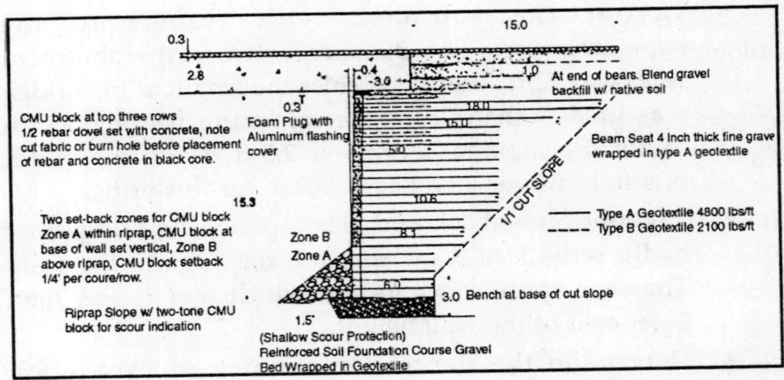

Fig. 7.22 Illustration. Reinforcement Schedule and RSF Dimensions for Bowman Road Bridge.

- *Add a bearing reinforcement zone underneath the bridge seat*: The primary reinforcement spacing was 8 inches at the wall face. The spacing of the bearing reinforcement bed was 4 inches, half of the primary spacing. The length of the bearing reinforcement bed was 5 ft. The depth of the bearing reinforcement bed would be determined when the internal stability analysis was conducted (step 7). At a minimum, however, there would be five intermediate layers between the primary reinforcement layers (at 8–inch spacing) in the bearing reinforcement zone.
- *Blend the reinforcement layers in the integration zone to create a smooth transition*: Additional work was needed to integrate the substructure with the superstructure within the integration zone at the approach. There were three layers of wrapped geotextile reinforcement spaced at 0.9 ft. This is described.

CALCULATE LOADS

The applicable surcharges and loads associated with the structure were a combination of vertical and lateral components. The vertical components include the surcharges due to the DL (superstructure and road base from the integrated approach) and the LL (superstructure and roadway), along with the weight of the GRS abutment. The lateral earth pressure due to the retained backfill, shown in table, were also considered. Lateral loads resulting from the DL and LL were calculated separately during the external and internal stability calculations performed in step 6 and step.

Table. Loads and Surcharges for Bowman Road Bridge.

Property	Notation	Measurment	Equation
Bridge DL	q_b	2,600 lb/ft²	Given
Bridge LL	q_{LL}	1,400 lb/ft²	Given
Roadway LL	q_t	298 lb/ft²	$q_t = h_{eq}\gamma_b$ $h_{eq} = 2.48\,ft$
Road base DL	q_{rb}	385 lb/ft²	$q_{rb} = h_{rb}\gamma_{rb}$ $h_{rb} = 2.75\,ft$
Weight of GRS abutment	W	9,257 lb/ft	$W = BH\gamma_r$
Weight of RSF	W_{RSF}	1,575 lb/ft	$W_{RSF} = B_{RSF}D_{RSF}\gamma_{rb}$
Weight of facing blocks	W_{face}	768 lb/ft	$W_{face} = N_{block}\dfrac{W_{block}}{L_{block}}$
Lateral load 'retained backfill,	F_b	5,258 lb/ft	$F_b = \dfrac{1}{2}\gamma bH^2K_{ab}$ $K_{ab} = \dfrac{1+\sin\phi_b}{1-\sin\phi_b} = 0.361$

Note that the weight of the GRS abutment was calculated with *B* equal to the shortest reinforcement layer not including the width of the wall face. This is a conservative assumption to simplify hand calculations. Several software programmes are available that can account for the varying shape due to different reinforcement lengths along the height of the abutment. The weight of the facing blocks (W_{face}) is the weight of an individual CMU block (42 lb) divided by the length of the block (15.625 inches), multiplied by the total number of blocks in a single column.

CONDUCT AN EXTERNAL STABILITY ANALYSIS

Direct Sliding

The driving forces on the GRS abutment include the lateral

forces due to the retained backfill, the road base, and the traffic surcharge.

The force due to the backfill is calculated in equation.

$$F_b = \frac{1}{2}\gamma_b K_{ab} H^2 = \frac{1}{2}(120)(0.361)(15.58)^2 = 5258\,lb/ft$$

The lateral force due to the road base and traffic surcharges are calculated in equation, respectively.

$$F_{rb} = q_{rb} K_{ab} H = 385(0.361)(15.58) = 2165\,lb/ft$$

$$F_t = q_t K_{ab} H = 298(0.361)(15.58) = 1676\,lb/ft$$

The total driving force (F_n) is then calculated in equation.

$$F_n = F_b + F_{rb} + F_t = 5258 + 2165 + 1676 = 9099\,lb/ft$$

The resisting force (R_n) is calculated according to equation. The total resisting weight (W_t) includes the weight of GRS plus the weight of the bridge beam plus the weight of the road base over the GRS abutment. Since the live loads are not permanent, they cannot be counted as a resisting force. Total resisting weight (W_t) is calculated in equation.

$$W_t = W + q_b b + q_{rb} b_{rb} = 9257 + 2600(4) + 385(0.7) = 19927\,lb/ft$$

The friction force (μ) is equal to tan Φ_{crit}. The interface friction angle between the reinforced fill and the geotextile was measured at 39 degrees by conducting an interface direct shear test. The resisting force (R_n) calculation is shown in equation.

$$R_n = W_t \mu = 19927 \tan(39) = 16137\,lb/ft$$

The factor of safety against direct sliding (FS_{slide}) is calculated in equation to make sure it is greater than1.5.

$$FS_{slide} = \frac{R_n}{F_n} = \frac{16137}{9099} = 1.8 \geq 1.5$$

Bearing Capacity

Before calculating the applied vertical bearing pressure, the eccentricity of the resulting force at the base of the wall must first be calculated using equation.

The moments are calculated around the center of the base of the RSF. The driving moments (calculated as a counterclockwise moment, include the lateral force due to the retained backfill, the road base DL) and the roadway LL. The calculation is shown in equation.

$$\sum M_D = F_b\left(\frac{H}{3}\right) + F_{rb}\left(\frac{H}{2}\right) + F_t\left(\frac{H}{2}\right)$$
$$= 5258\left(\frac{15.58}{3}\right) + 2165\left(\frac{15.58}{2}\right) + 1676\left(\frac{15.58}{2}\right)$$
$$= 57228 \text{ ft lb / ft}$$

The resisting moments (calculated as a clockwise moment) include the vertical force due to the bridge and road base DLs and the bridge and roadway LLs. The weight of the GRS abutment is also included as a resisting moment. This calculation is shown in equation.

$$\sum M_R = (q_b b + q_{LL} b)\left[\left(\frac{b}{2} + a\right) - \left(\frac{B_{RSF}}{2} - x_{RSF} - b_{block}\right)\right]$$
$$+ (q_t b_{rb,t} + q_{rb} b_{rb,t})\left(\frac{B_{RSF}}{2} - \frac{b_{rb}}{2}\right) + W\left(\frac{B_{RSF}}{2} - \frac{B}{2}\right)$$
$$= (2600 * 4 + 1400 * 4)\left[\left(\frac{4}{2} + 0.67\right) - \left(\frac{7.5}{2} - 1.5 - 0.64\right)\right]$$
$$+ (298 * 0.7 + 385 * 0.7)\left(\frac{7.5}{2} - \frac{0.7}{2}\right) + 9059\left(\frac{7.5}{2} - \frac{5.4}{2}\right) = 28098 \text{ ft lb / ft}$$

The total vertical load is equal to the sum of the weight of the GRS abutment, the weight of the RSF, and the load due to the DLs (bridge and road base) and LLs (bridge and roadway). This calculation is shown in equation.

$$\sum V = W + W_{RSF} + W_{face} + q_t b_t + q_{rb} b_{rb} + q_b b + q_{LL} b$$
$$= 9257 + 1575 + 768 + 298(0.7) + 385(0.7) + 2600(4) +$$
$$= 28078 \text{ lb / ft}$$

Thus, the eccentricity of the resulting force at the base of the RSF is calculated in equation.

$$e_{B,n} = \frac{\sum M_D - \sum M_R}{\sum V} = \frac{57228 - 28098}{28078} = 1.04$$

The applied vertical pressure is then calculated in equation.

$$\sigma_{v,base,n} = \frac{\sum V}{B_{RSF} - 2e_{B,n}} = \frac{28078}{7.5 - 2(1.04)} = 5180 \frac{lb}{ft^2}$$

The bearing capacity is calculated in equation The bearing capacity factors N_c, N_y, and N_q were found using table for the foundation friction angle of 0 degrees.

$$q_n = c_f N_c + \frac{1}{2} B' \gamma_f N_\gamma + \gamma_f D_f N_q$$

$$= 4000(5.14) + \frac{1}{2}(7.5 - 2 * 0.94)(120)(0) +$$

$$120(1.5)(1.0) = 20740 \text{ psf}$$

The factor of safety against bearing capacity failure is calculated in equation to make sure it is greater than 2.5.

$$FS_{bearing} = \frac{q_n}{\sigma_{v,base,n}} = \frac{20740}{5180} = 4.0 \geq 2.5$$

GLOBAL STABILITY

Global and compound stability was checked using the software programme ReSSA. Fig. 7.20 is a screenshot of the global stability failure mode. The factor of safety was found to equal 6.6, much greater than the minimum requirement of 1.5. Global and compound stability were satisfied.

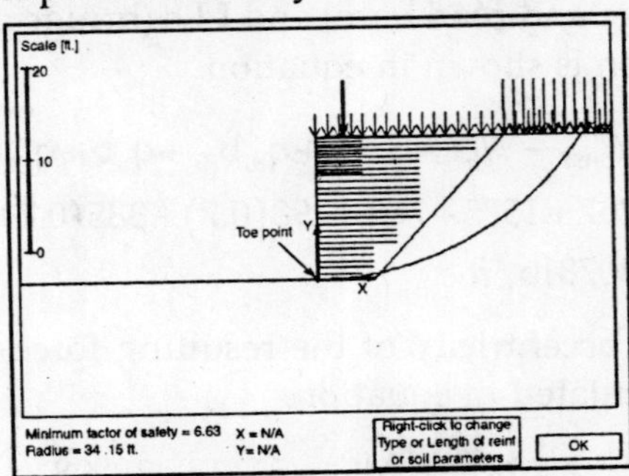

Fig. 7.23 Screenshot. ReSSA Results for Global Stability for Bowman Road Bridge.

ULTIMATE CAPACITY

The ultimate capacity of a GRS abutment can be determined using two different methods: empirical or analytical.

Empirical Method

The empirical method uses the load test results of a performance test on a GRS composite material identical (or very similar) to that used in the field. The ultimate capacity is found empirically as the stress at 5 per cent vertical strain from the stress–strain curve shown in figure. For this curve, the ultimate capacity ($q_{ult,emp}$) is 26 ksf.

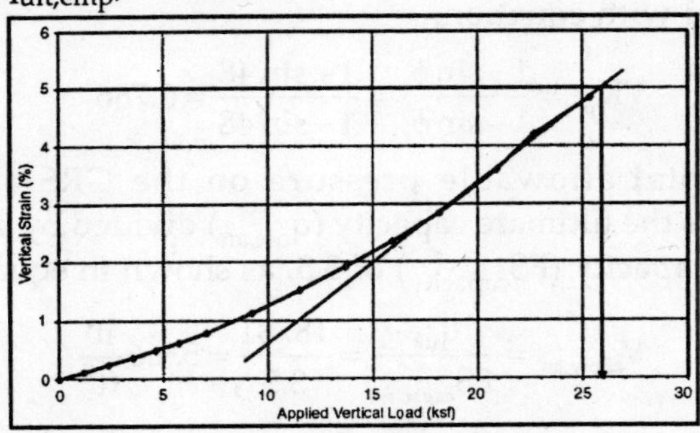

Fig. 7.24 Graph. Stress–strain curve for Bowman Road Bridge Showing Ultimate Capacity.

The total allowable pressure on the GRS abutment ($V_{allow,emp}$) is the ultimate capacity (q_{ult}) divided by a factor of safety for capacity ($FS_{capacity}$) of 3.5, as shown in equation.

$$V_{allow,emp} = \frac{q_{ult,emp}}{FS_{capacity}} = \frac{26000}{3.5} = 7429 \frac{lb}{ft^2}$$

The applied vertical stress ($V_{applied}$), which is equal to the unfactored sum of the vertical pressures on the bridge bearing area, must be less than $V_{allow,emp}$. This includes the DL from the bridge (q_b) and the LL due to the notional HL–93 load model (q_{LL}), as shown in equation.

$$V_{applied} = q_b + q_{LL} = 2600 + 1400 = 4000 \frac{lb}{ft^2} \leq V_{allow,emp}$$

Analytical Method

Alternatively, the ultimate capacity can be found analytically for a granular backfill. S_v is equal to 8 inches, d_{max} is equal to 0.5 inches, T_f is equal to 4800 lb/ft, and Φ_r is equal to 48 degrees. Although the spacing under the bridge bearing area was 4 inches, 8 inches was chosen in equation to be conservative.

$$q_{ult,an} = \left[0.7^{\left(\frac{Sv}{6d_{max}}\right)} \frac{T_f}{S_V}\right] K_{pr} = \left[0.7^{\left(\frac{8}{6(0.5)}\right)} \frac{4800}{0.67}\right] 6.786 = 18781\,\text{lb}/\text{ft}^2$$

The passive earth pressure for the reinforced fill was determined with equation.

$$K_{pr} = \frac{1+\sin\phi_r}{1-\sin\phi_r} = \frac{1+\sin 48}{1-\sin 48} = 6.786$$

The total allowable pressure on the GRS abutment ($V_{allow,an}$) is the ultimate capacity ($q_{ult,an}$) divided by a factor of safety for capacity ($FS_{capacity}$) of 3.5, as shown in equation.

$$V_{allow,an} = \frac{q_{ult}}{FS_{capacity}} = \frac{18781}{3.5} = 5366\frac{\text{lb}}{\text{ft}^2}$$

The applied vertical stress ($V_{applied}$), which is equal to the unfactored sum of the vertical pressures on the bridge bearing area, must be less than V_{allow}. This includes the DL from the bridge (q_b) and the LL due to trucks (q_{LL}). Applied vertical stress is calculated in equation.

$$V_{applied} = q_b + q_{LL} = 2600 + 1400 = 4000\frac{\text{lb}}{\text{ft}^2} \leq V_{allow,an}$$

DEFORMATIONS

Vertical

The vertical strain is estimated by using figure, as illustrated in figure for the bridge DL (q_b) of 2,600 psf. The vertical strain is therefore about 0.3 per cent–under the tolerable limit of 0.5 per cent. The road base surcharge is not included since it does not act over the same location.

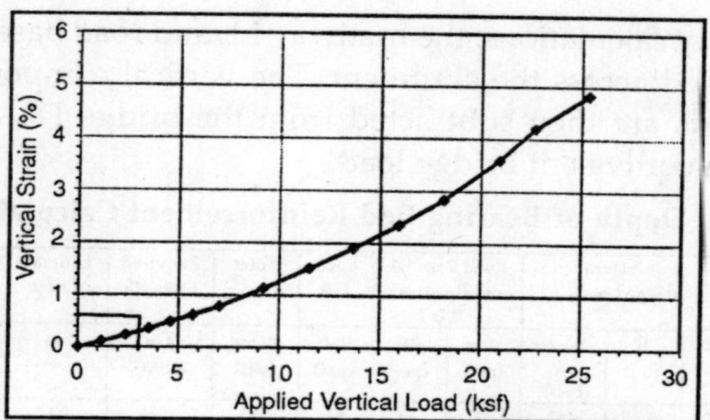

Fig. 7.25 Graph. Vertical Strain for Bowman Road Bridge.

The vertical deformation is the product of the vertical strain and the height of the GRS abutment (including the clear space distance), as shown in equation.

$$D_v = \varepsilon_v H = 0.003(15.58) = 0.047\text{ft}$$

Lateral: The lateral strain and deformation are found in following equation.

$$\varepsilon_L = 2\varepsilon_v = 2(0.3\%) = 0.6\%$$

$$D_L = \frac{2D_v}{H}(b + a_b) = \frac{2(0.047)}{15.85}(4 + 0.67) = 0.028\text{ft}$$

REQUIRED REINFORCEMENT STRENGTH

The strength of the reinforcement used at Bowman Road Bridge was 4,800 lb/ft. Applying a factor of safety of 3.5, the allowable reinforcement strength is 1,371 lb/ft. According to the manufacturer, $T_{@\varepsilon}$=2%is equal to 1,370 lb/ft.

The maximum required reinforcement strength is found as a function of depth, as shown in equation.

$$T_{req} = \left[\frac{\sigma_h - \sigma_c}{0.7^{\left(\frac{Sv}{6d_{max}}\right)}}\right] S_v$$

The lateral stress (σh) is a combination of the lateral stresses due to the road base DL (σh,rb), the roadway LL (σh,t), the GRS reinforced soil (σh,W), and an equivalent bridge load ($\sigma_{h\ bridge}$).

To simplify calculations, the roadway LL and road base DL can be extended across the abutment. The vertical components of these loads are then subtracted from the bridge DL and LL, giving an equivalent bridge load.

Table. Depth of Bearing Bed Reinforcement Calculations

Distance from top of wall	Equivalent Bridge Load			Road Base DL and Roadway LL		GRS Fill	Total	Required Strength	Ultimate Check	2 Per cent Check
z (ft)	α	β	$\sigma_{h,bridge,eq}$ (psf)	$\sigma_{h,rb}$ (psf)	$\sigma_{h,t}$ (psf)	$\sigma_{h,W}$ (psf)	$\sigma_{h,total}$ (psf)	T_{req} (lb/ft)	$T_{req} > T_{allow}$	$T_{req} > T_{\varepsilon_s=2\%}$
0.7	2.50	–1.25	482	57	44	11	593	1024	NO	NO
1.3	1.97	–0.98	449	57	44	22	572	987	NO	NO
2.0	1.57	–0.79	400	57	44	32	533	920	NO	NO
2.7	1.29	–0.64	350	57	44	43	493	852	NO	NO
3.3	1.08	–0.54	305	57	44	54	460	794	NO	NO
4.0	0.93	–0.46	269	57	44	65	434	749	NO	NO
4.7	0.81	–0.40	239	57	44	76	415	716	NO	NO
5.3	0.72	–0.36	214	57	44	86	401	692	NO	NO
6.0	0.64	–0.32	193	57	44	97	391	675	NO	NO
6.7	0.58	–0.29	176	57	44	108	385	664	NO	NO
7.3	0.53	–0.27	162	57	44	119	381	658	NO	NO
8.0	0.49	–0.24	149	57	44	130	380	655	NO	NO
8.7	0.45	–0.23	139	57	44	140	380	655	NO	NO
9.3	0.42	–0.21	129	57	44	151	381	658	NO	NO
10.0	0.39	–0.20	121	57	44	162	384	663	NO	NO
10.7	0.37	–0.19	114	57	44	173	388	669	NO	NO
11.3	0.35	–0.17	108	57	44	184	392	676	NO	NO
12.0	0.33	–0.17	102	57	44	195	397	685	NO	NO
12.7	0.31	–0.16	97	57	44	205	403	695	NO	NO
13.3	0.30	–0.15	92	57	44	216	409	705	NO	NO
14.0	0.28	–0.14	88	57	44	227	415	717	NO	NO
14.7	0.27	–0.14	84	57	44	238	422	729	NO	NO

The lateral stresses due to the equivalent bridge load are then calculated according to Boussinesq theory. The lateral stress is calculated for each depth of interest (each layer of reinforcement). All lateral stresses are calculated and shown in table. An example calculation for the required reinforcement strength at a depth (z) of 5.3 ft, or the eighth reinforcement layer from the top, is presented here.

First, the lateral pressure is found in equation. Remember, the location of interest is directly under the centerline of the bridge load where x = 0.5b =0.5(4ft) = 2 ft.

$$\sigma_h = \sigma_{h,W} + \sigma_{h,bridge,eq} + \sigma_{h,rb} + \sigma_{h,t} = 86 + 214 + 57 + 44 = 401 \frac{lb}{ft^2}$$

The calculation of each aspect of the lateral pressure is shown in equation through equation.

$$\sigma_{h,W} = \gamma_r z K_{ar} = 110(5.3)\left(\frac{1-\sin(48°)}{1+\sin(48°)}\right) = 110(5.3)(0.147) = 86\frac{lb}{ft^2}$$

$$\sigma_{h,bridge,eq} = \frac{(q_b + q_{LL}) - (q_{rb} + q_t)}{\pi}\left[\alpha_b + \sin(\alpha_b)\cos(\alpha_b - 2\beta_b)\right]K_{ar}$$

$$= \frac{(2600+1400)-(385+298)}{\pi}[0.72rad$$

$$+\sin(0.72rad)\cos(0.72rad + 2 * -0.36rad)]0.147 = 214\frac{lb}{ft^2}$$

$$\sigma_{h,rb} = q_{rb}K_{ar} = 385(0.147) = 57\frac{lb}{ft^2}$$

$$\sigma_{h,t} = q_t K_{ar} = 298(0.147) = 44\frac{lb}{ft^2}$$

The values for α and β are found in equation.

$$\alpha_b = \tan^{-1}\left(\frac{b}{2z}\right) - \beta_b = \tan^{-1}\left(\frac{4}{2(5.3)}\right) - (-20.7°) = 41.3° = 0.72rad$$

$$\beta_b = \tan^{-1}\left(\frac{-b}{2z}\right) = \tan^{-1}\left(\frac{-4}{2(5.3)}\right) = -20.7° = -0.36rad$$

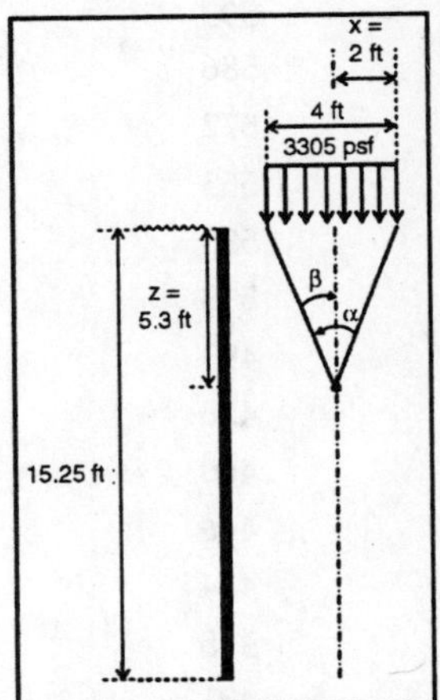

Fig. 7.26 Illustration. Lateral Pressure Due to the Bridge Load.

Based on table, the required reinforcement strength does not exceed the allowable strength or the strength at 2 per cent at any reinforcement layer. Therefore, no bearing bed reinforcement is needed; however, the minimum requirement is that the bearing bed reinforcement should extend through five courses of blocks. In actuality, six courses of block were chosen to extend the bearing reinforcement bed in this case 'to a depth of 4 ft below the top of the wall.

This was chosen to be conservative since this was the first bridge built with GRS technology. Applying 4-inch spacing to the top six courses of blocks and 8-inch spacing for the remaining height of the wall, the required reinforcement strength was found. The maximum required reinforcement is 716 lb/ft, which is less than the factored reinforcement strength of 1,371 lb/ft and the reinforcement strength at 2 per cent. There should, therefore, be no issues with reinforcement strength in the abutment.

Table. Required Reinforcement Along Height of Wall.

z (ft)	$\sigma_{h,total}$ (psf)	T_{req} (lb/ft)
0.3	594	319
0.7	593	318
1.0	586	314
1.3	572	307
1.7	553	297
2.0	533	286
2.3	513	275
2.7	493	265
3.0	476	255
3.3	460	247
3.7	446	239
4.0	434	233
4.7	415	716
5.3	401	692
6.0	391	675
6.7	385	664

7.3	381	658
8.0	380	655
8.7	380	655
9.3	381	658
10.0	384	663
10.7	388	669
11.3	392	676
12.0	397	685
12.7	403	695
13.3	409	705
14.0	415	717
14.7	422	729

IMPLEMENT DESIGN DETAILS

All design details were considered. Since it was a skewed bridge, the bearing area of 3 ft was maintained along the length of the face wall. The bearing bed reinforcement schedule was also maintained across the abutment face due to the superelevation, as shown in fig. 7.27.

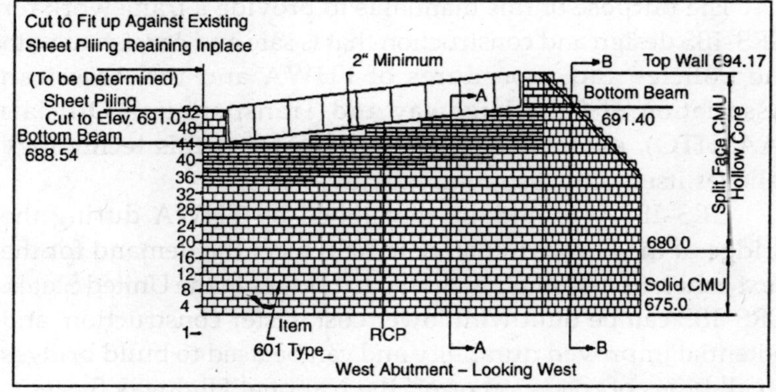

Fig. 7.27 Illustration. Secondary Reinforcement for Superelevation at Bowman Road Bridge.

FINALIZE MATERIAL QUANTITIES AND LAYOUT

The amount of reinforcement necessary is based on the

reinforcement schedule. Reinforcement material came in 12– to 18–ft–wide rolls. The number of facing blocks was determined from the height and length of the abutment and wing walls. The amount of backfill required was determined in a similar fashion. Once final quantities are established, it is a good rule of thumb to order at least 10 per cent more to account for unforeseen conditions.

GEOSYNTHETIC REINFORCED SOIL INTEGRATED BRIDGE SYSTEM

INTRODUCTION

The Geosynthetic Reinforced Soil (GRS) Integrated Bridge System (IBS) provides an economical solution to accelerated bridge construction. Employing this technology will help agencies save both time and money in planning and executing projects. This interim implementation manual and its companion document were developed to assist deployment of this promising technology as part of the Federal Highway Administration's (FHWA) Every Day Counts initiative.

The purpose of this manual is to provide a framework for GRS–IBS design and construction that is safe and consistent with the policies and procedures of FHWA and the American Association of State Highway and Transportation Officials (AASHTO), except where the behaviour of this technology relieves itself of those requirements.

GRS–IBS was initially developed by FHWA during the Bridge of the Future initiative to help meet the demand for the next generation of small, single span bridges in the United States. GRS–IBS can be built with lower cost, faster construction, and potential improved durability and can be used to build bridges on all types of roads, on or off the National Highway System.

GRS–IBS is a fast, cost–effective method of bridge support that blends the roadway into the superstructure to create a jointless interface between the bridge and the approach. It consists of three main components: the reinforced soil foundation (RSF), the abutment, and the integrated approach.

The RSF is composed of granular fill material that is compacted and encapsulated with a geotextile fabric. It provides embedment and increases the bearing width and capacity of the GRS abutment.

It also prevents water from infiltrating underneath and into the GRS mass from a river or stream crossing. This method of using geosynthetic fabrics to reinforce foundations is a proven alternative to deep foundations on loose granular soils, soft fine–grained soils, and soft organic soils. The abutment uses alternating layers of compacted fill and closely spaced geosynthetic reinforcement to provide support for the bridge, which is placed directly on the GRS abutment without a joint and without cast–in–place (CIP) concrete. GRS is also used to construct an integrated approach to transition to the superstructure. This bridge system therefore alleviates the "bump at the bridge" problem caused by differential settlement between bridge abutments and approach roadways.

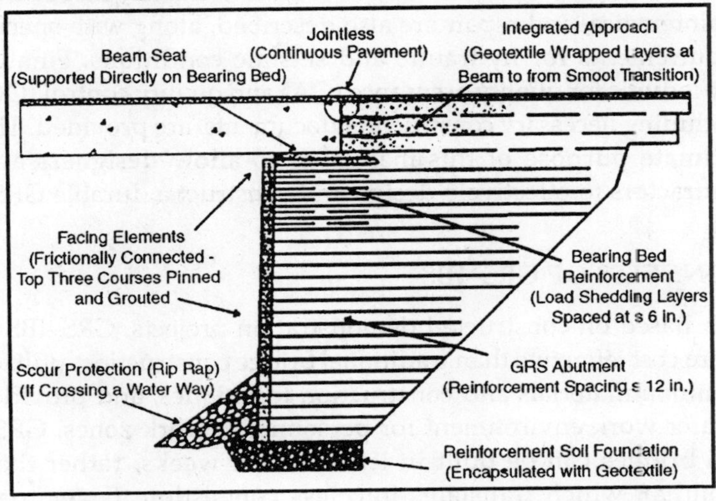

Fig. 7.28 Illustration. Typical GRS–IBS Cross Section.

The riding surface of GRS–IBS can be maintained as if it is part of the roadway pavement. No special attention to joints or the bridge deck is required. Unlike a traditional integral abutment, IBS is unique in its use of GRS to support the superstructure. This method of accelerated bridge construction is as easy as 1–2–3: (1) a row of facing blocks, (2) a layer of compacted granular fill, and (3) a layer of geosynthetic reinforcement. The 1–2–3 process is repeated until the required abutment height is reached.

GRS–IBS has many other distinct and innovative qualities. GRS technology is extremely durable and can perform well in earthquakes if constructed as outlined in this manual. GRS abutments can be built with readily available material using common construction equipment without the need for highly skilled labour. Construction of the abutment is contained within its footprint for a reduction of environmental impact as well as a reduced work zone. Additional benefits are convenience and design flexibility, as GRS–IBS can be built in variable weather conditions and can be adapted easily in the case of unforeseen site conditions. This manual addresses the design and construction of GRS–IBS. In–service performance, inspection, maintenance, and repair are also described, along with special requirements for hydraulic and seismic conditions. Finally, procedures for quality assurance (QA) and quality control (QC) 'including necessary construction documents, are provided. The ultimate purpose of this manual is to allow designers and contractors to effectively design and construct a durable GRS–IBS.

BENEFITS OF GRS–IBS

Based on constructed demonstration projects, GRS–IBS is more cost–effective than traditional bridge construction, utilizes common materials and construction techniques, and provides a safer work environment for personnel in work zones. GRS–IBS bridges can be built in less time (in weeks, rather than months), which translates into less congestion; fewer road closures, disruptions, and shutdowns around work zones; and lower materials and labour costs. The method of construction is such that the abutments are built from the inside out, reducing the exposure of personnel to potential roadside hazards. In addition, the technology is environmentally sensitive and results in minimal environmental impacts. The technology produces a reduced construction and carbon footprint, eliminates the need for installation of a deep foundation or CIP concrete, and can be adapted to fit the site–specific environmental needs.

The cost to build a GRS–IBS bridge is potentially 25–60 per cent less than traditional methods, depending on the standard of construction. The savings is attributable to the simplicity and flexibility of the design, speed of construction (which is less dependent on weather conditions than CIP abutments), use of readily available materials and equipment, and elimination of the deep foundation and other construction details associated with the approach way to the bridge.

Furthermore, this method has the potential for reduced maintenance costs because it eliminates the bump at the end of the bridge, creating a smoother and safer transition. Also, the application of GRS technology in other facets of earthwork (walls) culverts, foundations, slope stability, rock fall barriers, etc., has the potential to result in significant cost savings and more effective use of transportation funding.

In summary, the benefits of GRS–IBS include the following:

- Reduced construction time.
- 25–30 per cent lower cost than standard pile cap abutments on deep foundations with 2:1 slopes for off–system bridges.
- 50–60 per cent lower cost than standard department of transportation bridges.
- Construction that is less dependent on weather conditions.
- Flexible design that is easily field–modified for unforeseen site conditions.
- Easier maintenance due to fewer parts.
- Construction with common equipment and materials.
- Better quality control.

Index

U

V

W